My Cognitive autoMOBILE Life

Sebastian Wedeniwski • Stephen Perun

My Cognitive autoMOBILE Life

Digital Divorce from a Cognitive Personal Assistant

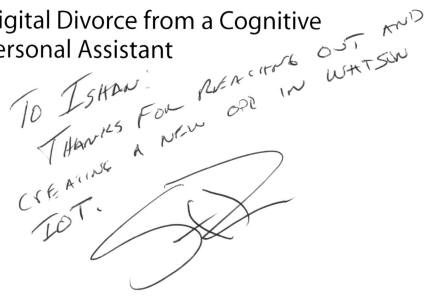

To ISHAN:
THANKS FOR REACHING OUT AND
CREATING A NEW OPP IN WATSON
IOT.

Springer Vieweg

Dr. Sebastian Wedeniwski
Standard Chartered Bank
Singapore, Singapore

Stephen Perun
International Business Machines
North Augusta, South Carolina

ISBN 978-3-662-54676-5   ISBN 978-3-662-54677-2   (eBook)
https://doi.org/10.1007/978-3-662-54677-2

Library of Congress Control Number: 2017957059

Springer Vieweg

Printed on acid-free paper

Proofreader: Zastrow information development GmbH
This Springer Vieweg imprint is published by Springer Nature
The registered company is Springer-Verlag GmbH, DE
The registered company address is: Heidelberger Platz 3, 14197 Berlin, Germany

Dedicated to our e-motion family who keep us continuously adapting to change.

Foreword by Helmuth Ritzer

Our relationship with cars is a highly emotional, almost an irrational one. Back in the 1950s and 1960s, the car was a sign of freedom, wealth and status. For many of my generation, owning a car was considered a major milestone in our life. When the oil crisis happened in the 1970s, I was a little kid and I remember asking myself if I would still be able to drive my own car.

Fast-forward more than 30 years. Meanwhile working in the automotive industry, I had another eye-opening moment during the financial crisis in 2008. For the first time in my career, I experienced a 30–40% decrease of car sales. What was typically a one-digit fluctuation in worldwide car sales completely changed from 1 day to the next. Even though the automotive industry wasn't part of the crisis, the car had lost its importance for many people who were impacted by the crisis. For many people, cars simply were no longer at the top of their mind and we finally had to realize that the automotive industry and the car as we know it was no longer a constant in our economy, something that used to sell itself no matter what.

In hindsight, what a coincidence that right at this time I got involved in a project, which was known as "City Smart". It was in 2008 when a group of people within Daimler AG were pulled together with the task of reinventing mobility. Today, we all know the product as "car2go", being the world's first free-floating carsharing service. It was for the first time that an OEM, in this case Daimler, thought about business models beyond pure car sales. For me, car2go, like many other services that came afterwards, demonstrates that there is an alternative model to car ownership. For many of the more than two million car2go users worldwide, the car is no longer this highly emotional thing that reflects your personality and which you still own, even though you hardly use it for more than 5% of the time. But this is just one of the change vectors.

The automotive industry is currently challenged in many ways. New players from the IT world are entering the market, e-mobility is considerably changing the value chain of the car, new clean air regulations and the infamous Dieselgate are heavily impacting the perception of the car and advances in AI and robotics will eventually make cars possible that no longer require a driver. The automotive industry is in a perfect storm, a storm that will change the face of the industry.

But rather than seeing these changes as a threat to an industry that is over 130 years old, there are huge opportunities out there, which will make possible new products, better user experiences and ultimately an opportunity to reinvent the car as something that will transition the AUTOmobile into something the authors call the autoMOBILE.

But no matter what the OEMs have done over the last decade to adapt the traditional business model towards this new reality, it is still a very car-centric one. For obvious reasons, each of them still puts the car at the center of their business; each of them wants to build their own ecosystem around the automobile, ignoring that ecosystems tend to be rather few and not many. What automakers need to realize is that they need to answer the mobility need rather than giving a specific answer, the car.

This is where the authors come in with their unique background both in the computer and the automotive industry. They bring together the latest advancements in machine learning, artificial intelligence, the IoT, cloud computing and the app economy with people's changing expectations of mobility and life in general. They describe how cars have been evolving over six generations from a complex but rather mechanically engineered device into a cognitive vehicle (third generation) that already leverages parts of the technology that is out there. But there are two more generations still to go until we can experience a vehicle that is fully embedded into our lives and contributes more to them than it generates issues for us and our environment.

It is a must-read for many of those engineering-driven decision-makers who rather than copy those troublemakers from the West Coast want to create their own way of evolving the automobile into a device that will be around for another 130 years at least instead. The chances are good, as the automobile has always been something very personal and private, something that will be key in the digital world in future as well.

VP Connected Vehicle Services   Helmuth Ritzer
HERE Technologies
Berlin, Germany
July 2017

Foreword by Michael Klingler

Technology is again transforming our way of living. One hundred thirty-two years after Gottlieb Daimler and Carl Benz invented the first car and 76 years after Konrad Zuse build the first computer, we are now part of the next huge transformation. The third generation of computing meets the next generation of mobility—two of the most important inventions and industries in our daily lives.

Some people believe that cognitive computing and artificial intelligence represent the third era of computing. We started from computers that could tabulate sums (1900s) to programmable systems (1950s), and now come to cognitive systems.

The way of being mobile is also changing to a new level; with autonomous and emission-free driving, we will see completely different usage and business models.

But there arise questions. What is intelligence in computers or even cars? How safe are autonomous vehicles? Even the attribute artificial brings up many different associations. It raises fears of intelligent cyborgs and recalls images of terror from science-fiction movies. It forces the question of whether our highest good, the soul and intelligence of a human being, is something that we should try to model or even reconstruct in machines.

I believe so, if we follow a basic principle that says that cognitive computing or AI systems always support people and let human beings make the final call. Then we are really beginning an exciting new way of how people and machines could interact and offer us new potentials and opportunities for our daily lives. Cognitive systems will forever change the way people interact with machines, help people extend their expertise across any domain of knowledge and make complex decisions involving extraordinary volumes of fast moving big data possible.

This technology will enable our *cars to become the third most important place besides our home and our office for managing our lives!*

With this new book, Sebastian is once more proving his brilliant sense for the correlation of new technologies and the transformation impact on the automotive industry. I've known Sebastian for 17 years and since our first meeting I've enjoyed collaborating and inspiring each other during many overnight sessions in Stuttgart, in different IBM labs and at many technology conferences in Las Vegas.

Sebastian is one of those people who bring real value to the automotive industry with their commitment. Thank you!

Now excuse me while I get back to the hard work of making this change happen.

Mercedes Benz R&D – Connected Cars   Michael Klingler
Stuttgart, Germany
July 2017

Acknowledgements

One of the greatest miracles of this world is seeing children grow up to be personalities with fantastic talents—thanks to my son Alexander for the fabulous illustrations. I am very proud of you! And thanks to my friend Stephen for all the inspiration and dedication.
Sebastian

She is my everything. Without her love and support, none of this would have been possible. Thank you, Sheila. Now to the person who made this book possible, Sebastian. Thanks for Japan and for challenging me to write this book with you.
Stephen

Contents

List of Abbreviations

ADAS	Advanced Driver Assistance Systems
AI	Artificial Intelligence
API	Application Programming Interface
BOM	Bill of Materials
CAN	Controller Area Network
CB radio	Citizen Band Radio
CEO	Chief Executive Officer
CPA	Cognitive Personal Assistant
CRM	Customer Relationship Management
DTC	Diagnostic Trouble Code
ECU	Electronic Control Unit
ESB	Enterprise Service Bus
GPS	Global Positioning System
HMI	Human Machine Interface
IoT	Internet of Things
IT	Information Technology
IVI	In-Vehicle Infotainment
IVR	Interactive Voice Response
KPI	Key Performance Indicator
M2M	Machine-to-Machine
NFC	Near Field Communication
NGTP	Next Generation Telematics Patterns
NLP	Natural Language Processing
NLU	Natural Language Understanding
OEM	Original Equipment Manufacturer
OTA	Over-the-Air
PAYD	Pay As You Drive
POI	Point of Interest
POS	Point of Sale
REST	Representational State Transfer
ROI	Return on Investment
SAE	Society of Automotive Engineers
SDL	Smart Device Link

SIM	Subscriber Identity Module
SMS	Short Message Service
SOA	Service-Oriented Architecture
STT	Speech-to-Text
TCU	Telematics Control Unit
TSP	Telematics Service Provider
TTS	Text-to-Speech
V2I	Vehicle-to-Infrastructure
V2V	Vehicle-to-Vehicle
VIN	Vehicle Identification Number
WAP	Wireless Application Protocol

Introduction

Humans are lazy, well, most of the time anyway, and we are not able to adapt to change. So, are we losing our intelligence?

This laziness and push to higher productivity promotes spaces, places, and devices to become more and more intelligent and interactive to assist us personally as we move through our daily life. Mirrors are no longer simple surfaces reflecting ourselves at the start of our day. Cars are no longer moving us impersonally from A to B. Cameras are no longer just capturing pictures of our day, and overall we do not need to go to the bank or even think about banking to pay for our needs in our *e-motion* day.

► But how comfortable are you when your surroundings are more intelligent than you?

Laziness and comfort are going hand in hand to incorporate our urge for assistance in our surroundings defined by intelligent spaces, places, and devices. But how personal and intelligent should the centralized assistance be? Centralizing all surrounding distributed intelligence will lead to designing a *personal assistant* that will go far beyond a chatbot [23]. It will lead to a broad discussion about where we will like or not like to have a digital assistant that is more or even less intelligent than we are. Nevertheless, it will get extremely personal if your personal assistant (see Fig. 1.1) predicts and acts on anything before you even think about it.

Twenty years ago—in a smartphone-free world—driving was about horsepower, torque, style and comfort. Those are the things you referenced when you said you loved your car, and "had to have" it. In about 20 years, the majority of locomotion using a kind of autonomous vehicle will be in what we can call robot cars. A car designer may say that a robot car should look like a car, but the design will evolve with the integration of many contextual services based on user input, location awareness, and the ability to access a variety of online information. These intelligent vehicles will be service robots adapting to any shape you will imagine. But what

© Springer-Verlag GmbH Germany 2017
S. Wedeniwski, S. Perun, *My Cognitive autoMOBILE Life*,
https://doi.org/10.1007/978-3-662-54677-2_1

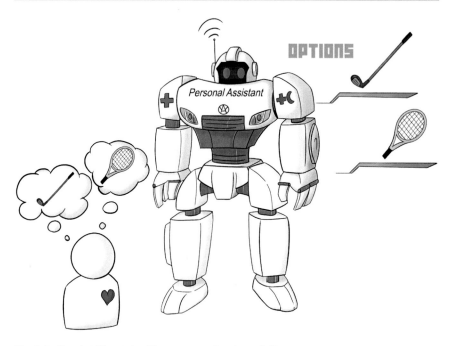

Fig. 1.1 How intelligent should your personal assistant be?

shape and how much intelligence will you allow your personal assistant to have, and will you again love it and trust it with your life? Will that love prefer "being" to "having"?

▶   Do we LOVE our cars?

We used to love our cars—it was a common phrase. Car manufacturers even had marketing campaigns that focused on just engine performance and emotional hardware designs [16]. This is clearly a physical world, with physical emotional attraction. Now consider how strongly air transport is regulated, with the result being that all planes appear very similar. Autonomously driving cars will also become more strongly regulated. Then the emotional attraction will be focused mostly in the interior. Moreover, the industry is transforming and people are changing so that the strongest, fastest or shrillest car may no longer be the most prestigious or bring the most meaning to our life. We hear statistics that most millennials in urban areas do not even want to own a vehicle, and use Uber and other forms of convenient transportation [8]. Time is becoming an ever more precious value in increasingly hectic times, which is being directly addressed by new mobility service providers. This cannot be good for car manufacturers. So can the brand now be the new *e-motion* in this mobility revolution [26]? Ford may think so, with their mobile application Ford Pass.[1] We do not even have to own a Ford vehicle to use it. Mercedes must think so. For example, when we're in Tokyo, we do not need a car, but we can still stop at the Mercedes-Benz Connection[2] for a coffee or tea. These are just a few examples of OEMs' thinking about this new *e-motion*, this new connection with their customers. Now how can OEMs start to make this connection in the vehicle or in this new world of mobility? Certainly, one way is through personalization and a learning system to understand the consumer. The idea of a *cognitive vehicle* that is more than just a physical engine, but a mobility experience personalized for me.

One example is when we jump in our car on Monday morning. Our vehicle immediately asks if we want our standard order placed at the coffee shop or if we want to use today's special offer of 10% off a new hazelnut coffee, since our profile shows a preference for flavored coffee. And based on traffic, it will be ready for pickup in 5 min. Our vehicle knows where we work, identifies our route, checks traffic and confirms we have no issues making it to work safely and in time for our first appointment. In summary, the car knows all our intentions when we are getting into it. Now this is a customer experience verses using our smartphone to open the application to order coffee, another application to check our route and traffic, or having to wait while it is integrated into our vehicle after which we have to select from our favorites near our work location.

► So, will we LOVE intelligent cars?

A *cognitive vehicle* —not only a physical but a personalized mobility experience managed by a *cognitive personal assistant*—which we cannot sell with just with a fancy vehicle engine name. Imagine a cognitive life, a similar experience when we get to the airport, pick up our rental car, and arrive at the hotel. The foundation to all of this starts with our identity, how we manage it, how we control and trust it, and all cognitive technologies and systems using it to create a more personalized experience.

Clearly, the automotive manufacturers are facing the market shift from an organization-centered to an individual-centered economy. More drastically, manufacturers will not survive the next 20 years if their integration competency stays on designing the body without value-adding brain-inspired systems. Technologies centered on *artificial intelligence* to mimic human intelligence have evolved quickly in the last few years. The software-defined life cycle of these technologies is so fast that not all these intelligent capabilities can simply be attached to a safe and reliable hardware-defined vehicle. The market will not wait for these manufacturers. Developer ecosystems are already quickly integrating various cognitive services into devices, spaces and physical objects, independent of the automotive industry.

---

[1] "FordPass – A Smarter Way To Move" http://www.fordpass.com. Accessed: July 26, 2017.

[2] "Mercedes-Benz Connection" http://www.mercedes-benz-connection.com/tokyo/ Accessed: July 26, 2017.

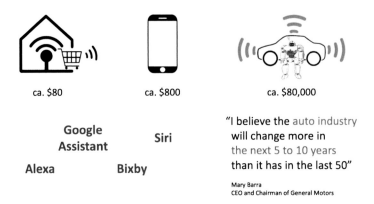

ca. $80   ca. $800   ca. $80,000

Google
Assistant   Siri

Alexa   Bixby

"I believe the auto industry
will change more in
the next 5 to 10 years
than it has in the last 50"

Mary Barra
CEO and Chairman of General Motors

Fig. 1.2 How to model a CPA? A CPA needs to care about safety, trust, transparency, multiple occupants/cognitive spaces, attention of the driver, mood of the driver, and connected brand lifestyle in an autoMOBILE context

In this book, we will use the following definition for a cognitive personal assistant:

> A *Cognitive Personal Assistant* (CPA) is a computer program that acts as an agent for an individual user to perform tasks or deliver services. These tasks or services are based, for example, on user input, location awareness, or information from a variety of online sources such as traffic conditions, news, weather, user schedules, etc. Current well-known examples of such a CPA are Apple Siri, Amazon Alexa, Google Assistant, IBM Watson, and Microsoft Cortana.

Soon, we assume we will see these CPAs at the level of an *autoMOBILE* and not just as part of a smartphone device (see Fig. 1.2). These personalized cognitive systems will not only provide insights and enable decision-makers, but will also act with confidence. These systems should provide full transparency and a unique personal ledger where all kind of data, preferences, and the software bill of materials of the service robot are securely managed in a blockchain that results in the journey of the cognitive autoMOBILE life that we will go into more in-depth and detail in this book.

1.1   Digitalization

The vehicle of the digital future will be created either by the automotive industry or by others who currently have better access to digital natives and hyper-connected generations. This new automotive paradigm is already here—but who will shape it?

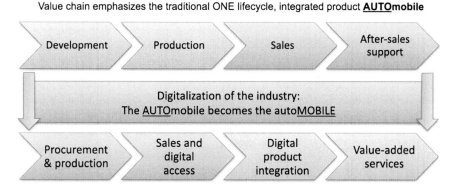

Value chain emphasizes the traditional ONE lifecycle, integrated product **AUTO**mobile

| Development | Production | Sales | After-sales support |

Digitalization of the industry:
The **AUTO**mobile becomes the auto**MOBILE**

| Procurement & production | Sales and digital access | Digital product integration | Value-added services |

Post-selling product-service system **auto<u>MOBILE</u>** to emphasize the mobility business model

Fig. 1.3 The changing automotive value chain will move the focus from before selling the integrated product AUTOmobile to an after-sale product-service system autoMOBILE

The future of mobility trends resulting from digitization, connectivity, personalization and data insights are fundamentally transforming the automotive industry in the coming years. In particular, large traditional vehicle manufacturers will have to develop new business models more radically and implement them much faster through new critical business skills. These new skills will go beyond the traditional automotive domains. The traditional competencies pertain to the AUTO component of the term "automobile", which are common primarily at the beginning of the automotive value chain, i.e. in the research and development, procurement, and production. In the future, the leading players in this disruptive market will have to put much more emphasis on MOBILE, the second component of the term "automobile". This requires investments in new digital technologies, new business skills, and partnerships with companies from sectors that have so far been unfamiliar to the automotive industry. In Fig. 1.3 we illustrate this changing value chain from the automotive to the mobility industry; more detailed information can be found in [26].

1.1.1 What Exactly Is the Product-Service System autoMOBILE?

The value chain creating the *autoMOBILE* underscores the mobility revolution with new business skills that are reshaping the vehicle of the digital future. This means the digitization of the industry requires a shift in core business skills from before to after the standard product sales model; highlighted in gray in Fig. 1.3. Instead of producing an integrated multi-feature product by the best engineers, new abilities are becoming crucial to build, buy and collaborate on a product-services system providing convenient mobility products. By moving the business skills related to system integration from development and production to a digital-oriented product

How to physically build a reliable and safe vehicle?	How to update services over-the-air/on-the-go?	How to model a cognitive and interactive personal assistant?
Hardware	**Software**	**Data**
Defined	**Defined**	**Defined**
HMI cockpit defined; cockpit based upon independently defined devices and individual part functions in an **AUTOmobile**	HMI is primarily a software-defined integration of multiple channels into one cockpit system; software controls holistically the moving **AUTOMOBILE**	Personalized HMI & ADAS is a semantic integration of data-defined services related to Spaces, Places and Devices to create the **autoMOBILE** as a service robot

Fig. 1.4 Different vehicle platform approaches today and in the future from a hardware, software and data-defined perspective

integration and changing value-added services, a software-defined vehicle platform needs to be newly designed that cannot be just attached to a safe and reliable hardware-defined vehicle. For example, a manufacturing-type digital copy based on a blockchain needs to be designed in a fast evolving software-defined life cycle besides the existing vehicle life cycle based on the well-known manufacturing approach build to order or build to stock. However, the personalization aspect of the autoMOBILE goes beyond software, because it is primarily defined by data. Hence, the CPA is getting a personalized human machine interface (HMI) that optimally manages all contextual services provided by the autoMOBILE. Figure 1.4 shows how a personalized experience is driving the future design of the vehicle platform from a data and software- defined perspective, in addition to the hardware-defined vehicle life cycle, which leads to a new relationship between driver and vehicle in [26].

1.1.2 Re-thinking Mobility

One significant effect of digitization is that the traditional market giants in the automotive industry will no longer be unaffected. New competitors or companies providing mobility services from other industries will be entering into the auto-MOBILE business, e.g. [7]. The photography and music industries felt a similar digitization and painful transition many years ago. Some giants from both industries have completely disappeared from the market and some new entrants in these disrupted industries grew as the new giants leading the new digitized markets. In the meantime, the automotive industry is also feeling similar market disruption pressures. New technology companies have already discovered this new and changing mobility market, even though the automotive industry is relatively protected by its high level of industrialization and very structured and tight ecosystem. However, the progressing digitization and the focus on service are causing the once clearly defined automotive industry boundaries to blur and even disappear [27].

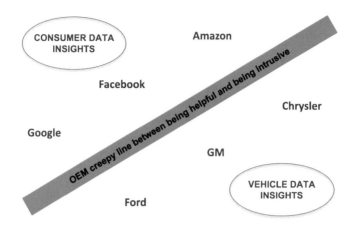

Fig. 1.5 The "creepy line" illustrates the breaking point between the two different business models where the company reputation and brand values are endangered if the line is crossed

How should the automotive industry react to these changes? There might be many different approaches, but in any case, it will not be enough to connect the vehicles to the Internet as a basic configuration and keep the traditional focus on manufacturing vehicles. Internet and software companies will be more radically innovating personalized customer experiences for all types of travel. In order to provide better experiences, even these companies will take the risk to leverage data that should not be used according to some regulated markets. Actually, most can "cross the creepy line"[3] which is nearly impossible for the OEMs with their current brand identity (see Fig. 1.5). Despite that, they need to get closer to this creepy line at least if they want to offer a personalized customer experience beyond selling an integrated product.

Tesla Motors, Local Motors, etc. plan for autonomous vehicles and Google, Uber, car2go, etc. show the changes in approach to the new mobility markets. An advantage of these "outsiders" is that they think about the mobility topic from the ground up, coming up with exciting ideas instead of bringing their expertise and technologies to the traditional automotive manufacturing market. Their business ideas are based on not just wanting to build better cars, but rather asking themselves the fundamental question of why there are cars at all and what the need for mobility means today. They have therefore not put forward the development and production of the vehicle as the center of their ideas.

These examples give us a good indication of where the industry is going in the future: The focus will not be on the production of AUTOmobiles, but on the provisioning of mobility services, digitization, connectivity and personalization. The data focus will be essential for this new business. In summary, the autoMOBILE

[3] "Schmidt: Google gets 'right up to the creepy line'" http://thehill.com/policy/technology/122121-schmidt-google-gets-right-up-to-the-creepy-line/ Accessed: July 26, 2017.

is changing to use new technologies and to integrate them into value-added business. The goal is to build new skills, develop new business models and implement them digitally—lean and agile principles.

1.1.3   Balancing Act Between the AUTOmobile and the autoMOBILE Business

A digital transformation is a complex challenge, which each company needs to solve individually because every company is unique. But there are many common core business skills in the automotive industry that can be structured in a reference model for today's AUTOmobile industry. More details can be found in [26]. Existing knowledge and expertise from the traditional AUTOmobile business must be transformed consistently using the new insights from customer data centered on the autoMOBILE business. Even if customers have reservations about sharing their data, they now expect personalized experiences, which makes a balancing act across the "creepy line" (see Fig. 1.5) necessary. Hence, a primary goal in reshaping the automotive business skills is to gain detailed consumer insights while still respecting consumer privacy. This will require a smart mix of consumer data and vehicle data insights.

The ability to gain detailed insights into the customer needs from their data will be one of the first key skills of the transformation road map of automotive companies. But companies who are primarily focusing only on digitization are missing that the digitization is only the foundation for new business. Technologies are becoming invisible but that is not the destination for the autoMOBILE business. The ability of a system to learn from customer experiences will be essential for the mobility of the future. Actually, we will discuss even further if learning should be put into any product [9], but next we want to introduce artificial intelligence.

1.2   Artificial Intelligence

After reading this section, you may be feeling a little overwhelmed by the complexity of the field of *Artificial Intelligence* (AI) and how the sheer breadth of theoretical concepts relates to the objectives of this book. Don't panic; give yourself time to become accustomed with these basics in order to understand the foundation throughout the entire journey to 'My Cognitive autoMOBILE Life'.

So slow down and let us start at the current market situation. We are seeing more and more examples of products released onto the market that are described by all kind of words like AI, augmented intelligence, cognitive system, machine learning, or natural language processing. Usually people are confused about the differences between them and in reality, we have to dig into the wording and definitions to understand what these systems really are and what they can actually do. Some key technologies and approaches to create the visionary AI goal are outlined in Fig. 1.7. Many AI approaches are deeply interconnected. For example, we distinguish between AI and robotics. In most cases, robotics is simply associated with the hardware container for the software and data-defined AI. In a biological analogy, AI is the brain and the robot is the body that receives an external stimulus. But robotics is closely associated with AI, therefore we list it as a "robot on its own" in Fig. 1.7, which describes a robot that is not (remotely) controlled by a human, which is basically the same as the description of the self-driving car in the automotive industry.

Common between all these words is the association with intelligence. But what is intelligence? Quite simply, human behavior can be intelligent, but on the other side, the complex behavior of an insect is not considered intelligent, so what is the difference? The physicist Stephen Hawking put it simply [4]:

Intelligence is the ability to adapt to change.

Hence, *intelligence* is the ability to adapt one's behavior to fit new circumstances, which is what a wasp cannot do.[4] This is a simple description of intelligence, but human intelligence that is discussed in literature has so many different dimensions that it is impossible to define it precisely by using just one sentence. For a more interesting and broader discussion on intelligence, we refer to the literature [24]. Interesting for our autoMOBILE focus is what Hawking said in this famous quote and how it can simplify our previous definition of the CPA and reflects how significant its role is by rethinking mobility in the direction of the "abilities to adapt to change" in this autoMOBILE world.

Now let us return to machines and more precisely, to when will we call them intelligent. The roots are going back to the year 1950, when Alan Turing asked the key question about AI [25]:

▶ Can machines think?

This was the beginning of the term AI and how it has evolved for the computer industry. However, early optimism concerning the attainment of human-level intelligence has given way to an appreciation for how profound and difficult the problem really is. We already realize how difficult it is just to define intelligence. Even when there is no general consensus on how to define or to understand "thinking", maybe a possible answer to "Can a machine think?" would be that a machine can perform an activity considered essential to humanity.

---

[4]"What is Artificial Intelligence?" http://www.alanturing.net/turing_archive/pages/reference%20articles/what%20is%20ai.html. Accessed: July 26, 2017.

Summarizing all, AI can be defined as:

Artificial Intelligence (AI) is a vision and a set of technologies and approaches that are needed by machines to achieve the problem-solving capabilities of humans.

It is common to introduce AI by using the following three types:

Narrow   are systems with the ability to complete pre-defined and limited tasks. Examples are search algorithms, spam filters, and recommendation engines.

General   are systems that achieve human-level performance across more than one domain (to mimic a human brain). An example is the CPA that should gain new skills in new domains without pre-programming or specific retraining.

Super   is a hypothetical agent with superior intelligence to all human brains across all domains, including creativity, social skills, and persuasion. Examples are still futuristic fantasies and fictional computers in movies like Deep Thought, the supercomputer charged with finding the answer to "the Ultimate Question of Life, the Universe, and Everything" in the science fiction comedy series "The Hitchhiker's Guide to the Galaxy" by Douglas Adams.

We will continue our discussion of AI using only the first two types. The third type is out of our scope, because creativity or persuasion is going far beyond our CPA and our cognitive autoMOBILE focus. Additionally, at the current time most of the groundbreaking developments are only occurring in the first two types.

Back to Turing. He did not just ask the AI question, by using principles of mathematical proof that could be processed by a machine, he showed that it is possible to make a machine "intelligent". In 1961 Marvin Minsky published in [13] how a machine can become intelligent through search, pattern recognition, learning, planning, and deduction. In the 1960s, Joseph Weizenbaum published an early *Natural Language Processing* computer program called Eliza. The concept was to simulate the responses a psychotherapist might have in a conversation. The Eliza program has no knowledge base or understanding of human behavior, instead the responses were achieved by pattern matching. In the 1980s the AI hype cycle reached a new public phase when the home computer became common and first versions of Eliza were released on the Commodore 64, (see Fig. 1.6).

These are just a few essential historic landmarks where much more development followed, see the literature [19]. Some key AI approaches are mapped in Fig. 1.7. For example, in the 1980s *expert systems* with their focus on knowledge and inference mechanisms were the most famous examples to achieve problem-solving

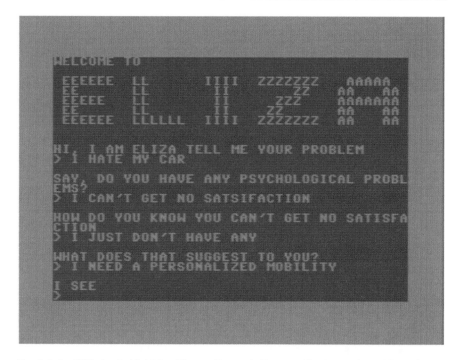

Fig. 1.6 In 1983 the Artificial Intelligence Research Group realized a version of Eliza for the home computer Commodore 64 which was almost identical to the original version. The computer program is just 121 lines of BASIC code [15]

abilities of humans by machines, i.e. pursuing the vision of AI. These systems were successful in narrow domains that could be structured, but had difficulty working with unstructured data and scaling like a human.

But is AI after more than six decades of research and development still only a scientific or technical topic today? Is it fear-generating hype, with the sensational headline: "AI can do your job faster and more accurately than you can!"? Or what is real and relevant for our industries?

Let us see by first looking at two recent and concrete examples that are in the AI field:

1. In February 2011 the *IBM Watson* computing system successfully demonstrated how to defeat the greatest contestants in America's quiz show "Jeopardy!".[5] This system was trained with pairs of questions and answers from past quiz shows to create a knowledge base for beating the record champion. The ability to top human players in a quiz show was important, but the real value of this

---

[5]"Computer Wins on 'Jeopardy!': Trivial, It is Not" http://www.nytimes.com/2011/02/17/science/17jeopardy-watson.html. Accessed: July 26, 2017.

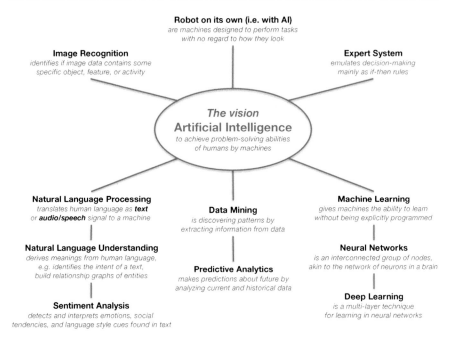

Fig. 1.7 Many different technologies and approaches are contributing to the AI vision. This map is not exhaustive, but instead outlines some of the key approaches

    system opened up a potentially new form of human-machine interaction and new applications, based on natural language processing

2. In March 2016 the *Google AlphaGo* computing system for the first time won a match against one of the top Go players without handicaps.[6] This system found its moves based on knowledge gathered by machine learning from previous moves and from extensive training, both from human and computer players.

    Beyond these two examples, to what extent has AI arrived in the industry and how far can it go? The answer lies in the ability of the industry, and how to leverage what is possible today:

1. Exascale computing speed,
2. big data,
3. and further developing AI technologies and approaches.

    Let us start with what is behind analytics and why it is essential for AI.

---

[6]"Google's A.I. Program AlphaGo Claims Second Victory Against 'Go' Champion" http://www.forbes.com/sites/parmyolson/2016/03/10/google-alphago-second-win-go-champion/. Accessed: July 26, 2017.

1.2.1 Analytics

Most people either have a love or hate relationship with mathematics. Let us open our mind to understanding analytics just a bit. Even we would be on the hate side because everything behind analytics is mathematics. A root of this love or hate situation is that mathematicians must intuitively understand and speak the language of business persons because all must have a common agenda:

Analytics is making better decisions in databased business.

The right analytics models can gain insight into the meanings of data and help the business to differentiate, but other models can have very little value if they are too abstractly developed and without business involvement. Therefore, mathematics is needed, but the agenda is bigger. Similar to having a pen or a keyboard in order to write a book for a specific audience—it is not enough to move with the pen from lines to letters, letters to words, words to sentences, and sentences to the final text of the book without adapting the mind to the bigger intended agenda. On the other side, the book stays vague inside some minds and cannot be communicated well and even successfully accomplished without the pen or keyboard.

In Fig. 1.7 we have seen that analytics is contributing to the AI vision that is the definition of it:

> *Analytics* is the discovery, interpretation, and communication of meaningful patterns in data.

There is extensive use of mathematics and statistics that use descriptive techniques and predictive models to gain valuable knowledge from data—data analysis. We will highlight a few different types of data analysis from trend analysis, spatial analysis, machine learning, and anomaly detection, which are common in the automotive industry.

Recently some companies have suggested skipping descriptive analytics and jumping right into AI. Bypassing the basics of analytics is not a shortcut to AI, it is like diving into AI, but struggling to embrace data. Remember, the same fuel powers AI as analytics—data. Without the transparency that analytics provides, it is difficult to judge the results of any AI system. AI cannot magically correct incomplete or poor-quality data. It is true that AI may be used to improve data quality, but it will not compensate for data that is fundamentally corrupt or unreliable. Overall, it is not a path to being successful with AI if you skip ahead without a solid analytics foundation.

For a successful path, we categorize analytics into the following five levels. This puts the terms in perspective to their competitive advantage and the value they bring

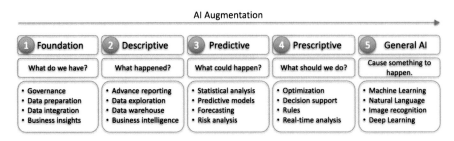

Fig. 1.8 Analytics in five levels to understand the value they bring to organizations

to a data-driven culture so that they can be fully embraced by augmenting AI, (see Fig. 1.8):

1. The first level is the foundation, the groundwork for data management, which leads to better understanding about what an organization has as data. Governance is the key to any good data management organization, because as the volume and types of data increases, a mistake early on will only be compounded. Integrating and bringing data together is an important step as well, because the first business insights were generated from basic reporting and viewing information.
2. The next level is descriptive, or the move to more advanced reporting and exploration of data to understand what is happening in an organization. Data warehouses and business intelligence created the ability to view and explore data in different ways, or views, to gain better insight. Data warehouses also combined data from many sources throughout the enterprise, bringing them together in a single location for more advanced reporting based on dimensions. A simple example of dimensions would be to look at customer data based on geography, a state, a county or just a city. In addition, early indications of alerts or warnings in reporting could be triggered to help identify problems sooner. No longer were reports static and only viewed upon request, but processed automatically.
3. The third level, augmenting AI, is predictive, the ability to estimate something before it happens, which is a very popular area in the automotive field and an area we will discuss in depth later. The key technologies related to statistical analysis and correlation are also used in the industry.
4. The fourth level is prescriptive, which we introduced as 'Narrow AI', not just looking at data at a given point in time, but in real time, where we can adjust and optimize a result using rules or basic decision-making techniques. Mostly used in manufacturing, it will become more popular as we move toward autonomous driving where data is very dynamic.
5. The final level is 'General AI' to achieve human-level performance across more than one knowledge domain. That will lead us to the next section introducing cognitive systems.

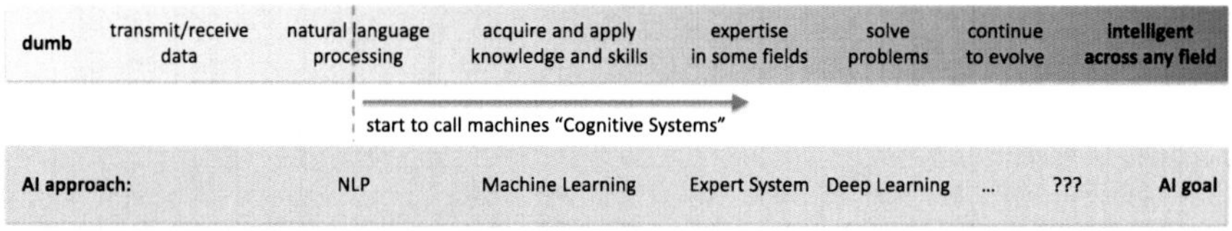

dumb	transmit/receive data	natural language processing	acquire and apply knowledge and skills	expertise in some fields	solve problems	continue to evolve	**intelligent across any field**

start to call machines "Cognitive Systems"

AI approach:		NLP	Machine Learning	Expert System	Deep Learning	...	???	AI goal

Fig. 1.9 A "speech and hearing" path of essential capabilities that need to be endowed to a machine to achieve the AI goal

1.2.2   Cognitive System

An intelligent machine cannot be an outcome of just one or even some of the AI approaches (e.g. mapped at Fig. 1.7). Let us try to describe a path for how to get to an intelligent machine, and for that it is easier to start with the contrast to what a dumb machine is. A dumb computing system is at best able to transmit or receive data without any independent processing capabilities. But when do we start to call a machine intelligent? For machines, it might be similar to human intelligence, but there is a wide gray area between what is dumb and what is intelligent. Let us draw a picture (see Fig. 1.9) to better understand a path of the essential capabilities that need to be endowed to a machine to achieve the AI goal. Figure 1.9 is a simplified but not exhaustive path to outline key capabilities because the human's ability to hear and understand natural language as a step into the cognitive systems definition is not enough to achieve the strong AI goal. Besides hearing, other human senses like sight, smell, taste, and touch also need to be considered, depending on the use cases. But the aspects mentioned in the gray area are not the primary focus of our discussion whether AI approaches are possible so a machine can get "a bad feeling about something" or how AI can be trained for e-motion and soft skills.

Back to Fig. 1.9. We see *Natural Language Processing* (NLP) as a fundamental building block on the "speech and hearing" path to building a cognitive system, because it translates human language to a machine language without manually programming the machine code. Now this type of human-like communication can lead to more complex conversations for acquiring and applying knowledge where we see *machine learning* approaches as essential to give cognitive systems the ability to learn without being explicitly programmed. Sophisticated modern learning techniques enable programs to generalize complex rules from data. Today, most types of learning are not automatic, which means that the training models are only as smart as the manually created training sets and the provided data. Identifying the right training data sets still takes manual work and adjustment, maybe just as much as much as programming used to take. However, once we get through all that effort we achieve a cognitive system that has expertise in some fields. It can act like a sophisticated *expert system* to answer most questions based on the learned rules for limited knowledge domains. The next stage for solving problems—not following just a rule-based model—is more complicated, because it goes beyond answering specific questions. An example of one problem to solve on the "speech and hearing" path is to explain the same thing in an appropriate way, but to different types of people. A recent breakthrough in machine learning called *deep learning* is driving the most success on that stage, especially incorporating multiple data sources to get more insights of the persons in the conversations. All further stages, where a cognitive system is continuously evolving or is an intelligent expert across any knowledge domain, are future theory and more hypothetical. Nevertheless, AI approaches will be developed further in the future and these new stages might be possible in the next few years.

In summary, we introduced the term *cognitive system* inside the gray area describing a path of approaches to accomplish the AI goal.

> A *cognitive system* is an IT system that organizes a set of AI technologies and approaches to assist in solving or to solve a variety of industrial and consumer-related tasks or problems.

By that definition, a cognitive system is not intended to achieve artificial super intelligence, for example, to be creative.

But even after more than six decades of AI research and development, we are still in the early days of these promising new technologies and different AI approaches. We only now have the necessary computer speed, architecture and data volumes to support the ideas. Most importantly, all these technologies and approaches are radically different from the programmable systems that have been produced by the IT industry for half a century and are primarily running everywhere today. These technologies build systems that are not explicitly programmed by instructions in computer languages, they learn—from their own experiences, their interactions with humans or other machines and build on the feedback and outcomes of their learnings.

So let us consider some additional key contributors to the AI vision of enabling cognitive systems.

1.2.3 Natural Language Processing

In Fig. 1.9 natural language processing (NLP) was illustrated as the entry into cognitive systems. Actually, we will have a more detailed introduction of why and how NLP defines the customer experience. As such,

> *natural language processing* (NLP) is the key process of explaining instructions to a machine, in the natural language, as used by humans.

NLP systems are usually divided into two fields: text and audio. Audio plays a significant role for NLP inside the vehicle. It is obvious that it is less dangerous to listen than to read the output from a mobile phone when driving a car.

But typically a driver is first signaled by an audio prompt that the in-car system is waiting for a voice command. Voice commands may be used to initiate phone calls or enter destinations for the navigation system. Hence, the first technology goal is to label the noise bursts before we enter the linguistically based understanding inside NLP. The following two systems are essential for speech-based communication:

1. *Speech to text* (STT) provides the ability to convert spoken word into text, also known as speech recognition. Acoustics models are at the core of STT implementations, in which the speech recordings and corresponding text are input to match the sounds that make up words statistically. Simply put, speech in and words out, but there is still a difference regarding how the speech should be transcribed into text. An example:

Raw text:	hello how are you today i'm frustrated i cant open my car
Simple formatting:	Hello, how are you today? I'm frustrated, I cannot open my car!
Speaker change:	Agent: Hello, how are you today?
	Customer: I'm frustrated, I cannot open my car!
Emotion/affect:	Agent: Hello, how are *you* today?
	Customer: I'm *frustrated*, I cannot open *my car*!

The biggest challenge is the noise inside the vehicle coming from different sources such as the engine, the environment (e.g. rain or wind), the infrastructure (traffic on the road), and even other passengers' speech.

2. *Text-to-speech* (TTS) provides the ability to convert text to a spoken word, also known as speech synthesis. This technology originally used smaller recorded sound pieces to put them together ultimately to form the synthesized sound. The original TTS systems normalize text, assign phonetic transcriptions to each word, and finally convert the symbolic linguistic representation into sound. The quality of TTS is judged by its similarity to the human voice and by its ability to be understood clearly. In 1992 the first voice navigation system with audio-guided functions was developed by Aisin AW and adopted for the Toyota model Celsior.[7] Only recently have deep learning techniques been introduced to mimic realistic human voices by sampling natural human speech and directly modeling waveforms.[8]

---

[7]"Car Navigation System" http://www.aisin-aw.co.jp/en/products/information/. Accessed: July 26, 2017.

[8]"WaveNet: A Generative Model for Raw Audio" http://deepmind.com/blog/wavenet-generative-model-raw-audio/. Accessed: July 26, 2017.

As part of the NLP challenge, Natural Language Understanding (NLU) enables computers to derive meanings from natural language input, because the same natural language constructions may bear different meanings in different contexts. In NLU, a model or set of words are needed to describe a specific domain. For example, to order food the system needs the basic understanding of meats, vegetables, etc. along with drinks, sizes and types. This understanding of the words is built into a corpus that is then used in NLP to understand a sentence of text.

Understanding human language is the key communication system to make direct interactions between humans and computers possible. This is going deeper into linguistics than Eliza (see Fig. 1.6) we introduced, which uses only pattern matching techniques. Imagine a computer that prompted a customer: "Would you like to pay a bill or transfer funds?". There is a good chance that the customer's response will contain either "pay a bill" or "transfer funds". They sound very different so that neither STT nor NLU would be necessary for these kinds of questions and answers. We summarize this as noise burst classification, but not as NLP. Our challenging goal is that a CPA or any other cognitive system communicate with the same fluency and articulation as humans, which goes beyond the current applications in search engines or question-answering systems. More detailed information related to NLP can be found in [17].

Let us be more specific with an example. What happens if we send the following text to a cognitive system:

I'm frustrated, I cannot open my car!

Firstly, using the Google Natural Language API for syntax analysis, we create a dependency parse tree for this sentence (see Fig. 1.10). Getting the structure of the sentence is the foundation for further analysis, but just the structure is not enough to understand the sentence. For that we will need other APIs. For example, Google provides an API to extract the object that correctly identifies a "car", (see Fig. 1.11). In addition, this API provides the information that the car is a consumer good, which is useful for branching off into a specific knowledge domain. In this way, we can

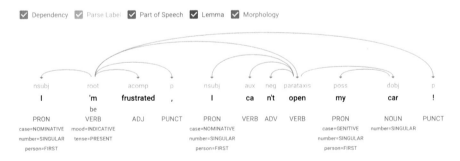

Fig. 1.10 Google Natural Language API for syntax analysis creates dependency parse trees for each sentence

language.googleapis.com/v1/documents:analyzeEntities

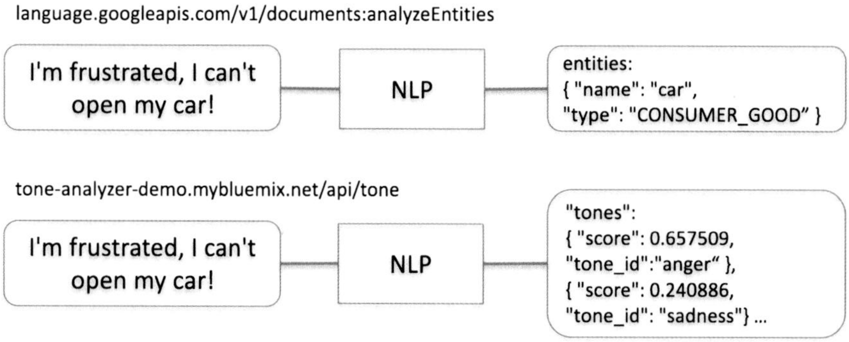

| I'm frustrated, I can't open my car! | NLP | entities: { "name": "car", "type": "CONSUMER_GOOD" } |

tone-analyzer-demo.mybluemix.net/api/tone

| I'm frustrated, I can't open my car! | NLP | "tones": { "score": 0.657509, "tone_id":"anger" }, { "score": 0.240886, "tone_id": "sadness"} ... |

Fig. 1.11 Google Natural Language API extracts the object of a free text and the IBM Tone Analyzer service detects the emotions found in text

further analyze the text structure to identify, for example, the intent "open" as well. There are many more such text analysis services, for example, the extraction of date, time, location, or address if context is provided within the text. Another important part of text analysis is identifying the emotional tone behind the text. For that we can use, for example, the IBM Watson Tone Analyzer service, where this text input will return "anger" as a likely emotional tone with a confidence of about 66%, (see Fig. 1.11).

The next step in NLP after analyzing the input text is the dialog, which models a conversation flow starting with a simple question or a series of questions, answers and potential responses. There are two types of dialogs to consider:

Linear dialog   has one aim, which is to collect the information necessary to complete the required action, e.g. find a cheap shared car.

Non-linear dialog   has several branches, depending on the user's answers, e.g. was the ride good?

In our example to build linear dialogs, a customer can request a car in many ways depending on how he words his request:

- "I need a car now."
- "Find a car to Tokyo Tower."
- "Looking for a car from my current location to Tokyo Tower in 5 minutes."

Now we are using the slot-filling feature of the Api.ai, where all these example requests can be listed in the "User says" section and all parameters used to perform a search in the "Action" section. The prompts are needed for the dialog to ask the user if a request does not address all required parameters for a search.

Contexts:	Users says			
	"I need a car @when"			
	"Find a car to @to"			
	"Find a car from @from to @to @when"			
Action:	find.car			
Required	Parameter name	Entity	Value	Prompts
Yes	destination	@to	$to	Where do you want to go?
Yes	depart	@when	$when	When?
Yes	starting point	@from	$from	Where do you want to start?

Now in this context we get back to the situation: "I'm frustrated, I cannot open the car!" That will start a more complex non-linear dialog to specify and we refer to the Api.ai dialog documentation.[9]

We selected Google, IBM, and Api.ai just as examples. Many others provide similar results. For example, Microsoft Language Understanding Intelligent Service, Facebook Wit.ai Bot Engine, Amazon Alexa Skill Set, Api.ai and a growing list of NLP startups, especially with an industry specialization focus. All of these providers have a good foundation on designing and building a chatbot [23].

But a human might identify more information out of this one sentence than an NLP alone can. For example, the "frustration" indicates the context of frequent trials to "open", which could mean the door or car key might be damaged or that the key is actually inside the locked car. This kind of hypothesis needs to be validated by further context, but getting to such context needs more than NLP, which lead us to machine learning.

1.2.4   Machine Learning

What do we think of first when we hear *machine learning*? Most people might picture a robot, but a robot without AI is just a hardware container that cannot act on its own. So, we introduce the following definition from the AI pioneer Arthur Samuel [21]:

> *Machine learning* is the AI approach that gives machines the ability to learn (from data) without being explicitly programmed.

Now what does the ability to learn from data really mean? Let us go back in computing history to when we had tabulating machines, a piece of specifically

---

[9]"Dialogs" http://api.ai/docs/dialogs. Accessed: July 26, 2017.

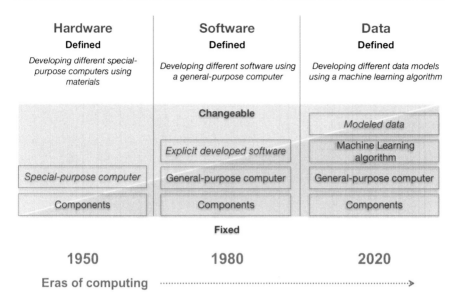

Fig. 1.12 Different IT computing system eras from the hardware, software and data perspective evolved in the last 70 years

developed hardware for solving a computation problem. Then came general-purpose computers, where software and programs were written to solve many types of problems. Now we have AI and models that use "general-purpose" algorithms and data to solve problems. We summarized this difference in Fig. 1.12, which relates also to the different vehicle platform approaches described in Fig. 1.4. Let us describe how a machine can apply learning to solve a problem. One example is effectively managing the high number of unwanted and irrelevant emails in our mailboxes. That would be a fundamental problem that a CPA has to solve so that it can focus on providing only the necessary information in an appropriate way for different types of people.

First, we want to solve the problem by developing a program as shown in Fig. 1.13. The traditional approach is to write a program manually using instructions of a machine language with the resulting program becoming a long list of complex rules. For example, the program will formulate rules for blocking certain addresses or finding some keywords and phrases that tend to come up a lot in the emails. That sounds pretty simple, but the problem of the anti-spam system is not to allow a lot of spam through and, on the other hand, not to reject any legitimate emails. The challenge is to tune the content based on statistical means correctly to the types of legitimate email that an individual gets; and these types are changing over time. Hence, what is good for one can be spam for the other and what is good today can be spam tomorrow. That all requires hand tuning of the rules, is hard to maintain and even rewriting the program is continuously required.

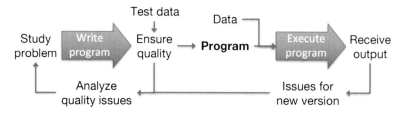

Fig. 1.13 Traditional approach for programming machine code to receive an output for a problem

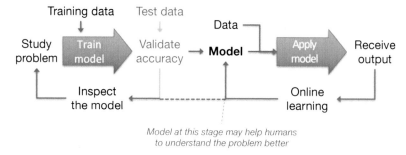

Fig. 1.14 The machine learning approach for achieving a continuously adapting model optimizing the output for a particular problem

Now let us look at how a machine learning approach to solving this problem is different. What's important is that this approach will not claim—to use our email filter example—to filter spam more accurately than the traditionally developed program. However, it shows how the filtering model can individually evolve and adapt easier to change—remember, how we have introduced "intelligence" at the beginning of Sect. 1.2 by quoting Hawking. A first look at this approach (see Fig. 1.14) reveals it to be more complicated than the traditional approach (see Fig. 1.13). Breaking down the two major phases of these two different approaches shows:

1. Instead of manually writing a computer program that has to be tested to ensure its quality, a model for the machine learning algorithm is trained to discover accurate relationships within a set of data and may be validated by a different set of test data.
2. Instead of executing a computer program on new data, a trained model is applied to a machine- learning algorithm with the data of new cases. However, the new data may self-modify the model to improve future outcomes further.

In summary, the traditional approach is driven by a program, the machine learning approach is data-driven by a model. The model represents the discovered relationships across different data types and their values—which are often called *features*.

Training as the First Phase

Let us start to understand better what the training in the first phase of Fig. 1.14 means. Actually, training is like learning with the intent to acquire knowledge, but with a focus on creating a model from scratch by using a (manually) created larger training set of data. That process can be considered batch learning. The resulting model can only be as good as the quality of the training set is and how representative the training is to generalize for new cases. Usually, the training takes a lot of time and computing resources, so it is typically done offline from the second phase. After the initial discovery of relationships between the features, a test set may be used for validating how accurate the discovered relationships are, but this validation is not possible for all types of trainings.

This first phase is different from the traditional programming approach. An explicitly programmed code will usually not be completely rewritten after identifying quality issues. However, in this batch learning, a new version of the model has to be trained from scratch on the full dataset, i.e. not just the new data, but also the old data.

Especially, in the training phase, machine learning is classified into two types:

> *Supervised learning* uses a labeled training set of pairs of input data and desired output values. In *unsupervised Learning*, the goal is to discover hidden structures in the training set of data without any pre-determined labels.

Actually, there are more types in the literature, *semi-supervised learning* which combines both types we previously defined, is between completely labeling training data and without labeling them at all. Another type is *reinforcement learning* which allows the machine or software agent to learn its behavior based on feedback from a specific context. Think of maximizing the performance of the CPA; reinforcement learning would, by trial and error, determine that certain assistance in a specific context would enhance the driving experience. However, reinforcement learning would be more relevant in the second phase than in the first phase with the focus on training.

Back to the two primary training types—supervised and unsupervised learning. There are dozens of algorithms on how to implement these learning approaches. Supervised learning is divided into two broad categories:

Classification   categorizes data into a predefined class specified by the label. For example, the CPA can be trained with sport categories of personalized interests in golf, running, cycling, and swimming.

Regression   is a continuous measure of the relation between the mean value of one variable and corresponding values of other variables. For example, the CPA can be trained with energy consumption and driving behavior for range prediction of an electric vehicle.

Unsupervised learning is also divided into two broad categories:

Clustering	is grouping—not predefined—data in such a way that data in the same cluster are more similar to each other. For example, the CPA might cluster the interests by topics or by occupants inside a vehicle.
Dimensionality reduction	simplifies data without losing too much information. For example, the CPA identifies that a shared car mileage closely correlates with its age, so they will merge them into one feature that represents the car's wear and tear.

By these broad categories, it is clear that unsupervised learning does not have the intent to build a perfect model, but often helps to understand the problem better. Clearly reducing dimensionality does lose some information, just like compressing an image to JPEG degrades its quality, but it reduces complexity and speeds up the training.

In Fig. 1.15 we illustrated these four different learning categories based on the example problem of how a machine can learn to filter spam emails, which is an essential responsibility of a CPA to focus on providing only the necessary information. A practical approach for supervised learning is to split the data of a training set into two thirds to train the model and one third to validate the accuracy of the outcome. This approach works because the training data is labeled, but this approach does not work for unsupervised learning. That is the reason why we have highlighted this step "validate accuracy" in gray in Fig. 1.14.

Applying Learnings as the Second (Repeating) Phase

In the first phase, there are dozens of learning algorithms to train a model, but the primary challenge is not to select the best learning algorithm for the known training data. It is more important to foresee which algorithm and model will predict the most accurate class or value of the new data in the second phase. Besides selecting the best learning algorithm with the outlook on the second phase, the most common issue is insufficient quality of training data. The learned model is then less likely to perform well on predictions for new data. Insufficient quality of data does not only mean errors, it happens also if the training data is noisy.

non-representative	for the new data, e.g. trained music recognition model based on music data from popular global artists, will not work well in Japan,
non-fitting	in the sense of too simple to generalize well, e.g. train life-satisfaction models based on earnings data,
unfocused	on relevant features, e.g. train car performance based on brand names.

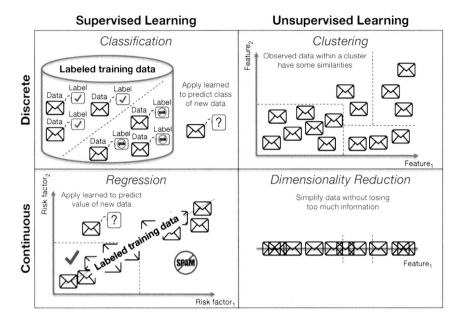

Fig. 1.15 Some important learning categories using the example of how a machine can learn to filter spam emails

In addition to the quality of data, millions of representative examples are needed for complex problems. Today, tremendous progress is happening in many areas like speech recognition (NLP audio) where millions of examples are generated by the mass of mobile phone consumers.[10]

Applying a good learning model is like executing a well-developed program, so the second phase is not too different from the explicit programming approach. It is only significantly different if we allow the model to self-modify itself by applying new data. This automatic adaptation to change is called *online learning*, which we can see in Fig. 1.14. This "automatic adaptation to change" loop makes this approach intelligent, see introduction of intelligence in Sect. 1.2. But then the approach will no longer be deterministic, i.e. generating the same output by applying the same input data. Furthermore, poor data quality in online learning gradually degrades the model, so it is a continuous challenge to rely on a stable outcome using machine learning.

There is a range of technologies and languages, e.g. Java, Python, and Scala, for developing core machine learning algorithms. But it is more common for AI applications to leverage pre-built machine learning frameworks, for example

---

[10]"Why Baidu's breakthrough on speech recognition may be a game changer" http://www.scmp.com/magazines/post-magazine/article/1925784/why-baidus-breakthrough-speech-recognition-may-be-game. Accessed: July 26, 2017.

Mahout[11] is popular on Apache Hadoop, but Apache Spark MLlib[12] has become a de facto standard.

Stepping Back

By now we know what high-level machine learning is and what its advantages and limitations are, especially regarding what we discussed about which loop makes the best approach intelligent by continuously adapting to change. The major limitation of machine learning to mimic human thought is that a model is needed as input that can be further self-modified, but the creativity to build a completely new model cannot be learned by this AI approach. Hence, it does not innovate or figure out which problems to solve, but it does discover patterns in large quantities of data to predict new values.

We have seen the difference—the traditional approach is driven by an explicitly written program, while the machine learning approach is data-driven by a model. That should provide us a better understanding of how to think about an approach to design the vehicle platform from a data-defined perspective (see Fig. 1.4), and we now know the relevance of a CPA to build the autoMOBILE.

Deep Learning

Deep learning is a class of machine learning techniques that can utilize either supervised or unsupervised algorithms, or both as semi-supervised. In particular, these techniques became popular because of the successful solutions for the three typical channels text, speech and vision. For the one specific problem of NLP, see Sect. 1.2.3.

Let us look into a deep learning application that reads someone's handwriting to understand the advantages of this learning technique. In Fig. 1.16 every digit—provided by the MNIST database[13]—is written in a slightly different way. By taking just one look at Fig. 1.16, we can easily recognize the context of similar rows with ascending numeric digits, i.e. every digit in the first column as a zero, every digit in the second column as a one, etc. Otherwise, we could recognize circles or the letter O in the first column, and slashes or backslashes in the second column if we did not have the broader context. Now assuming we know the input context of only Arabic numerals but without any ordering, then we can start to write an explicit program to identify the digits by rules to differentiate one digit from another. For example, the program might identify a zero if the image has only one single closed loop.

---

[11]"Apache Mahout" http://mahout.apache.org. Accessed: July 26, 2017.

[12]"Apache Spark MLlib" https://spark.apache.org/mllib/. Accessed: July 26, 2017.

[13]"THE MNIST DATABASE of handwritten digits" http://yann.lecun.com/exdb/mnist/. Accessed: July 26, 2017.

Fig. 1.16 Examples of handwritten numeric digits from a MNIST dataset

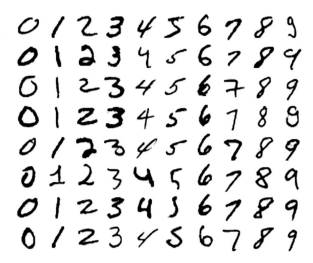

But the programming challenge is to decide the trade off when a loop is considered closed, e.g. see the zero in the third row in Fig. 1.16. The challenge behind this kind of programming will get larger if we have to differentiate an illegibly written zero from a six. Here we can measure the distance between the starting and ending point of the loop as what will work in many cases of scrawled zeros and sixes. But defining rules on how to distinguish single digits needs to be extended to cases like one and seven or four and nine or even eight and nine. If you look at the second row, you see that will become a real challenge for the program. At the end, this will result in developing many explicit rules for a machine to be able to distinguish what a human can with just one look.

Actually, handwriting is a much more complex problem. We have introduced the simple reading of Arabic numerals, where the identification scope of the ten numeric digits is quite narrow. After the previous introduction of machine learning, we know that there are different approaches to tackle these more complex problems. Instead of spending major efforts for a supervised training to improve reading someone's handwriting, the deep learning approach can find usable structures and representations through unsupervised training of a massive amount of unstructured input. What finally ends up in a semi-supervised training is starting with supervising the basics of the letters and then letting the system learn through the mass of handwriting.

But how is that possible and why should we use deep learning?

In 2006 a groundbreaking paper [5] reignited the artificial neural network field with multi-layered neural networks that are now more commonly referred to as *deep learning* in AI. The training of these networks, combined with the advances in computing power and the availability of mass amounts of data, outperformed the other machine learning techniques such as regression and clustering (see Fig. 1.15). So,

> *Deep learning* is a class of machine learning techniques that exploits many layers of non-linear information processing for supervised or unsupervised feature extraction and transformation, and for pattern analysis and classification.

Within the past few years, various descriptions have evolved from deep learning research that have widened the original scope, but most common among all definitions are two key aspects referring to [2]:

1. Models consisting of multiple layers or stages of non-linear information processing; and
2. Methods for supervised or unsupervised learning of feature representation at successively higher, more abstract layers.

Let us understand a key element of deep learning, the multi-layered architecture, by going back to the example of the handwritten numeric digits. The approach is to build up from basic objects into more complex representations through multiple layers of connected nodes. Instead of directly identifying the shape and entirety of a digit, the input layer starts with nodes, where each node responds to a specific kind of line. These processed inputs as pixels in an image are triggering the first hidden layer where basic features are recognized, such as diagonal, horizontal, and vertical lines. In Fig. 1.17 the nodes connected through different layers are shown as a graphic model. Exemplary, these inputs are data representing parts of the digit "4". We are talking about deep learning when multiple hidden layers exist and this is where most of it is happening to solve the problems. That means in our example, after processing the basic features of simple lines in the first hidden layer, more complex features like right, acute, or obtuse angles are recognized in the second hidden layer before we proceed to the outcome "4", which in our case consists of four right angles and two acute angles. In Fig. 1.17 the connections with a stronger weighting are in red to highlight the identification process of our example digit "4". But in Fig. 1.16 we can easily see that even a clearly written "4" can also appear to be made up of only horizontal and vertical lines or of only diagonal lines or even curves. Further hidden layers are becoming relevant to identify numbers with multiple digits in a specific context such as a currency. In our example, we have spent a lot of effort to identify useful features for the different layers, but the advantage of deep learning is to leverage the hidden layers to automate this

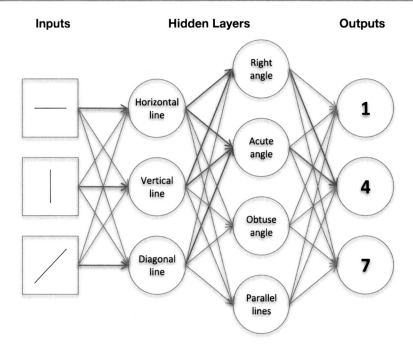

Fig. 1.17 An example of a four-layered architecture of deep learning connects nodes from pixels of lines into representations of numeric digits. Red are the exemplary weighted connections to output "4"

process, i.e. analyze features which the model automatically learns to extract from the unsupervised data.

In a real implementation, thousands of input nodes may be required to respond to different kinds of lines and curves representing a handwritten digit and to finally identify this digit successfully in the output layer. Actually, the graphic model is more wide than deep, because usually there are just a few hidden layers to process features but a wide range of nodes. Finally, we should mention that we only illustrated the simplest way to analyze a *feed-forward network* in Fig. 1.17 where the processing is only flowing from left to right and no connections between the nodes in the same layer exist.

These learning techniques therefore give a CPA the core capability to learn the individual text, voice and vision to ensure the best personalized customer experience. But these core capabilities need to be packaged as small, autonomous services to scale in many different situations without developing every possible combination of context. Microservices provide us the way to efficiently leverage this ecosystem of capabilities for interactions between the CPA and a human.

1.3   The Microservices Way

In the 1990s the Service-Oriented Architecture began to evolve for distributed systems, see the example [22]. Essentially,

> *Service-Oriented Architecture* (SOA) is a software design approach where collaborating services provide a set of capabilities. A service represents an encapsulated business functionality to provide a specific outcome to consumers.

The related buzzword 'service-orientated' promotes loose linking between services, which goes beyond modularizing an application. This means that separately maintained services communicate interoperably via calls across a distributed network, rather than method calls within a computing process boundary. Hence, SOA is independent of products and technologies.

So, how are microservices different or are they just a modern interpretation of SOA?

The definition of microservices is in a similar situation as SOA was 25 years ago. It is simply an overhyped word with ever-changing interpretations. However, let us focus more on the fundamental idea of minimalizing modular software development. Let us look at a set of cultural norms that were summarized by the experiences of leading developers of the Unix philosophy, see [20]:

* Write programs that do one thing and do it well.
* Write programs to work together.
* Write programs to handle text streams, because that is a universal interface.

The definition of Application Programming Interfaces has evolved through this modular software development.

> Generally speaking, an *Application Programming Interface* (API) is a clearly defined description of a software component to expose a set of data and functions so other software pieces can exchange information with it.

Building modular software applications without APIs is basically like using a conventional car without a gas tank. To a computer, the API is the same as what filling the gas tank is for a car, the interface to the outside world. Now elevate this API definition in line with the Unix philosophy to facilitate interactions between heterogeneous and distributed systems running across different computing

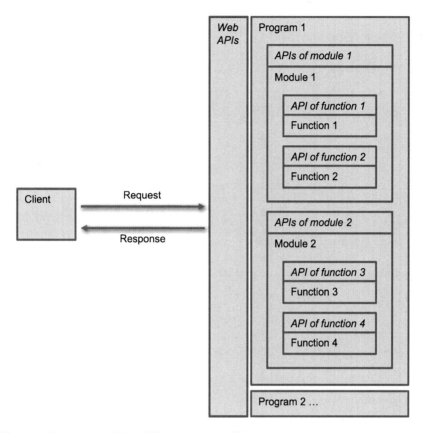

Fig. 1.18 Examples of APIs at different levels specifying the communication between software components

processes, specifically in the network that operates over the Internet. These different levels of APIs are exemplarily illustrated in Fig. 1.18.

Basically, *web APIs* allow software components to scale the access to data, services, or other resources over a connected network.

Hence, web APIs are directly listening and responding to running web client requests, for example, in a web browser. Over the last few years, the interoperable web API standard *Representational State Transfer* (REST) is gaining momentum and leads the way. REST is an style of architecture satisfying formal constraints. In particular, the application must be separated into a client-server model and the server must remain stateless. A web API conforming to the REST architectural style is a REST API. More detailed information regarding REST can be found in [11]. In the next chapter, we will use REST API examples for different platform generations to exchange information with vehicles by making use of the HTTP verbs to support all the CRUD operations:

POST	Creates a new resource.
GET	Retrieves a resource.
PATCH	Updates parts of an existing resource.
DELETE	Deletes a resource.

Example operations related to customer information are:

Action	URL	Description
GET	/account/{id}/address	List addresses by customer ID
POST	/account/{id}/address	Add new address to customer
PATCH	/account/{id}/address/{id}	Edit address by ID

It is important to realize that we can make the same call of the action PATCH repeatedly while producing the same result, but not of the action POST. We will not show complete REST API implementations for the platform discussions in next chapter. Our intention is to enable the microservices style of architecture to target the design of the platform for 'My Cognitive autoMOBILE Life'.

We will not design all APIs from scratch because there are already many well-established REST APIs available that provide capabilities and services in the cloud. For example, we will leverage existing REST APIs from:

- Google Cloud Platform
  http://cloud.google.com/products/
- IBM Bluemix
  http://console.bluemix.net/catalog/
- Amazon Web Services
  http://aws.amazon.com
- Microsoft Azure REST API Reference
  http://docs.microsoft.com/en-us/rest/api/

More specific REST APIs we will leverage in the scope of an autoMOBILE are under development, for example by HERE.[14]

We introduced SOA and API in today's open market, and to gain an understanding of REST in the context of the web APIs design. How are microservices now related to all that?

---

[14]"REST APIs from HERE Maps – HERE Developer" http://developer.here.com/develop/rest-apis. Accessed: July 26, 2017.

It is well known that the entire business of Uber and Netflix is built on the back of publishing APIs, e.g. Netflix Open Source Software Center.[15] But their approach is not associated with SOA even though Netflix was founded in 1997 at the time of SOA, but slightly before the definition of REST. So what is different?

Microservices are not about services and are not about being small, so it is not surprising there is not much clarity on what it means. Let us define how we see it:

> *Microservices* are a style of architecture for dividing the implementation of an application into a set of components, a focused piece of software completing one task only with an own independent life cycle—i.e. independently developed, deployed, upgraded—and commonly renders its business benefit as services via HTTP/REST.

This definition again follows the Unix philosophy, but the focus of microservices is on designing the endpoints of specific business functionalities and not on technology. Therefore, microservices can be developed in any programming language and are communicating via the language-neutral REST APIs.

In summary, SOA and microservices architecture share many of the same business-focused principles, but the business and technical motivations for microservices are different. SOA uses services to integrate different applications through orchestration, while microservices instead structure an application into lightweight services by avoiding the problems of tiered architectures. As a consequence of the broader scope, standards, and tooling that SOA had in the past, a microservice architecture intentionally focuses on solving a single problem with lightweight endpoints. We conclude this discussion with the following comparison:

Microservices	SOA
One application has multiple microservices	Integrates multiple applications
Focus on endpoints, not on pipes	Focus on ESB
REST, JSON etc	SOAP, XML, WSDL etc

For a more in-depth look at microservices, see the literature [14].

Reflecting this hype cycle of different styles of architecture, we should still be aware that most APIs today follow the traditional programming approach, (see Fig. 1.13). These interfaces are logically designed and linked with transaction processing systems. For example, a task could be "A pays B" where an API would be designed with the attributes from, to, amount, currency, and time when to execute the payment. These are fundamental APIs for building ecosystems, especially in the

---

[15]"Netflix Open Source Software Center" http://netflix.github.io. Accessed: July 26, 2017.

world of the *Internet of Things* (IoT). They also improve the customer experience through more flexibility, but APIs need to leverage more AI capabilities to enable the CPA to deliver the best customer experience. Just imagine, if the CPA would provide only an API "A complains about B", but that is not the customer experience we are aiming for.

1.4   Customer Experience

The era of the personalized customer "my way" has founded the customer experience revolution!

In the last few decades the development of digital capabilities and rising consumer expectations have resulted in the move from an organization-centered to an individual-centered economy, (see Fig. 1.19). Just think about that it was only 10 years ago that the first iPhone was released by Apple and how that has significantly redefined the customer's perceptions of the usage of mobile phones and how and when to get engaged. At that time, Steve Jobs said:

> You've got to start with the customer experience and work back toward the technology.

Nowadays, customers drive consumption and expect businesses to communicate in their language, whenever they want, however they like to. Simply said: "Know me!" That leads to the definition:

> *Customer experience* is the product of interactions between an organization and a customer over the duration of their relationship.

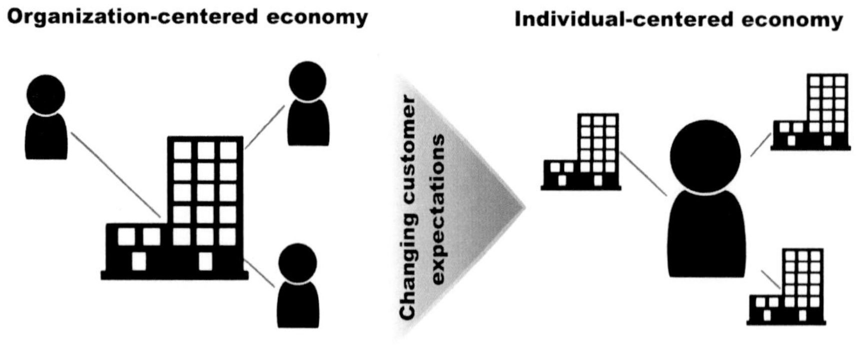

Organization-centered economy   **Individual-centered economy**

Fig. 1.19 Digital technologies are changing consumer expectations, transforming the organization-centered economy into an individual-centered economy, see [6]

However, one single positive interaction between the organization and customer does not make a good customer experience. The interactions need to be consistent and seamless throughout the whole relationship life cycle to achieve customer loyalty and retention. Moreover, all different channels of interactions need to be integrated for seamless customer experiences. It is easy to lose customers through only one poor interaction, but it is very hard and time consuming to win them back.

Now more than ever, communication preferences are driving customer experience expectations. From a human-machine communication perspective, NLP defines the future experience, (see Sect. 1.2.3). Only customer services that are easy, simple and efficient lead to a repeatable positive experience. But the challenge with today's AI solutions is that the human-machine communication is based on one-step recommendations, often called *next best action* to respond to the customer's needs during the interaction. One-step recommendations will usually not lead to the best long-term relationship and customer experience. It is more like playing chess where a combination of recommendations will lead to the best customer experience.

Let us ask ourselves what it means to be alive "my way". Is it to be able to focus on ourselves? Is it to be able to learn continuously? Or is it to seek giving back to communities? These kinds of questions lead us to different thinking.

First, we have to think outside the product, service or system to improve the flow of interactions within the individual-centered economy. That leads us to changing the design processes with the introduction of *design thinking* and the focus on rethinking and reimagining the customer experience, new ideas and emerging products. The fundamentals of design thinking are to how to best engage with customers.

> *Design thinking* is a human-centered approach that starts with customer empathy, and then matches what is technologically feasible and what a viable business strategy can convert into customer value and market opportunity.

In Fig. 1.20 we illustrate how the focus is shifting to integrating customer desirability with technological feasibility and business viability to deliver innovative solutions to transform the customer experience.[16] It is very important for a business to create a personalized experience. The process starts with the definition of a persona for our understanding of the customers, their behaviors and characteristics. An exercise on needs, pain points and the as-is process helps ensure we are looking at the problem from the customer's point of view. Next in the exploration phase, we look at potential to-be ideas. Not limiting ourselves to technical capabilities, but pushing towards what the customer would really want. We prioritize those ideas and put them into a story or journey map of how the customer would experience

---

[16]"Design Thinking—Thoughts by Tim Brown" http://designthinking.ideo.com. Accessed: July 26, 2017.

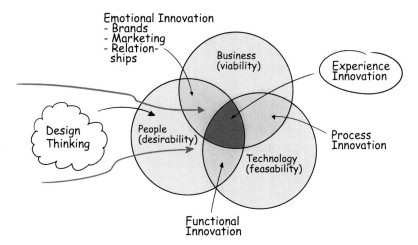

Fig. 1.20 Design thinking integrates what customers desire with technological feasibility and business viability to deliver innovative solutions to transform the customer experience (Source: IDEO)

that product throughout the day or a week. Next, we move into a prototyping phase, which can be an animation, movie, or even screen mock-ups or full-fledged prototypes.

This is where we start to be Agile[17] and choose to follow a set of values and principles, but the final step is feedback and reprioritization based on that feedback. Incorporating the products and this real data from the field can only cause greater improvements to the design. Agile philosophy with design thinking creates faster feedback cycles on this new methodology as well. Specifically when the changes can be done in software, the idea of releasing new features and products weekly is not unusual. Products can be released much faster and sometimes with the wow feature, but with a road map to quickly implement more and more features.

A short note on Agile: We do not want to get into a debate about software development frameworks—when to use certain ones or which is best. We will just briefly discuss the parts of the philosophy that work best with design thinking and when and how we explain solutions to non-technical lines of business. Agile with its more iterative and adaptive practice tends to work better when there is not a full set of known requirements up front. Figure 1.21 shows an incremental development if we were to approach a painting, and Fig. 1.22 shows the iterative development. The new focus and passion of design and development teams will have more flexibility to create and build along the way. An agile development best expressed by these two

---

[17]You cannot "do" Agile, you can only "be" Agile.

Fig. 1.21 Incremental development approach to a painting

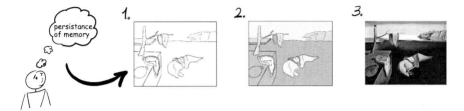

Fig. 1.22 Iterative development approach to a painting

painting approaches, which we think clearly shows the difference between iterative and incremental development.[18]

For further reading on design thinking, we refer to [1]. Let us see the customer experience, specifically around disrupting the automotive industry.

1.5 Personalized Mobility Experience

The *personalized mobility experience* is basically the customer experience while using modes of transportation to overcome physical distances. In 1886 the era of the AUTOmobile pioneered by Carl Friedrich Benz began to disrupt the personal relationship between humans and horses by enabling new mobility experiences through a "horseless carriage". Actually, that vehicle was a very expensive, rare and imperfect product without any evidence as to how useful it would be for the public. In August 1888, the personalized mobility experience started when Bertha Benz embarked on the world's first long-distance journey in automotive history. Her trip was key to understanding the potential for an individual mobility experience at a novel speed and range. That personalized mobility experience, defined by an uncertain and challenging venture, is different from the AUTOmobile driver experience promoted nowadays by car manufacturers—currently only the general

---

[18]"Don't Know What I Want, But I Know How to Get It" http://jpattonassociates.com/dont_know_what_i_want/. Accessed: July 26, 2017.

perception and emotional state of the driver.[19] This AUTOmobile promotion leads us to the beginning of the introduction (see page 2) when we asked the question:

▶   Do we LOVE our cars?

But buying a unique lovely AUTOmobile will be too narrow to associate it with a personalized mobility experience. Actually, reflect on Fig. 1.5 from a customer experience perspective and ask the following questions:

- Is buying a car a good customer experience today?
- Is car insurance a good customer experience today?
- Is car maintenance a good customer experience today?
- Is parking a good customer experience today?
- Is trading in a car a good customer experience today?

This is the entry for non-automotive industries to deliver the autoMOBILE as a broader personalized mobility experience to keep up with ever-changing demands to fit into an increasingly complex network of transportation options. Now compare the travel experience when Uber became popular, adapting new business models, which nowadays is similar to what Bertha Benz experienced on her venture without any digital capabilities 130 years ago. Hence, we define:

> *Personalized mobility experience* as traveling around in a way that is made for me.

In fact, the AUTOmobile meets the majority of user needs for individual travel, but the autoMOBILE has the goal of being adaptable to user needs, which makes it intelligent—see the quote by Stephen Hawking on page 9. The ability to adapt to interconnectedness is the essence of the creative disruption in the mobility industry ahead: between consumers and automakers; between consumers and vehicles; and among traditional and non-traditional participants in the industry ecosystem. That leads us to the cognitive vehicle, and to how consumers can access it in new ways of their digital lives.

---

[19]For example, "MAZDA – Be a driver." http://www.mazda.co.jp/beadriver/. Accessed: July 26, 2017.

1.6 Cognitive Vehicle

Based on the introduction of the cognitive system (see Sect. 1.2.2) combined with the need of a personalized mobility experience, we can now define what a *cognitive vehicle* is:

> A *cognitive vehicle* is a physical vehicle leveraging information and a set of AI technologies and approaches to assist users in and around the vehicle, related to the personalized mobility experience.

To understand this better conceptually, we start with a forward-thinking study "Automotive 2025: Industry without borders" [7]. That study is based on the automotive industry moving to engage with consumers, embrace mobility, and exploiting the ecosystem. The executive summary points out how the automotive industry was a closed ecosystem with little input from customers, and how with the introduction of the digital experience from other industries the automotive industry would need to change. This study shows that the dynamics of the consumer-vehicle-enterprise relationship are starting to change drastically as traditional industry boundaries disappear. Uber and car2go are great examples of these types of transformations as they see stronger market growth in mobility than their traditional core business where the customer experience is primarily defined by selling the AUTOmobile.

It is this kind of thinking that drives the IoT across all industries and pushes the automotive industry to think differently. The following six themes were defined in the study (see Fig. 1.23) that we associate with the foundations of the cognitive vehicle:

- Self-integrating,
- Self-configuring,
- Self-learning,
- Self-healing,
- Self-driving, and
- Self-socializing.

Therefore, the cognitive vehicle is centered on the concepts of taking care of its occupants, taking care of itself, and working with others. Let us understand these self-enabling themes when used as a foundation in defining the cognitive vehicle. First,

Self-enabling vehicles

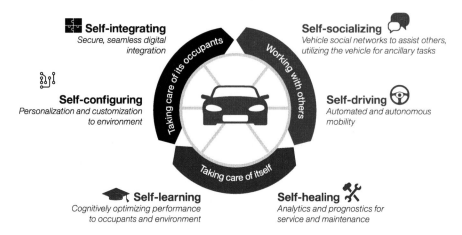

Self-integrating
Secure, seamless digital integration

Self-socializing
Vehicle social networks to assist others, utilizing the vehicle for ancillary tasks

Self-configuring
Personalization and customization to environment

Self-driving
Automated and autonomous mobility

Self-learning
Cognitively optimizing performance to occupants and environment

Self-healing
Analytics and prognostics for service and maintenance

IBM

Fig. 1.23 The foundations of the cognitive vehicle is self-enabling centered on the concepts of taking care of its occupants, taking care of itself, and working with others (Photo: IBM Institute for Business Value [7])

> *self-integrating* is a secure, seamless digital integration of the vehicle into the IoT. It enables information sharing by a vehicle with other parties, the IoT, and vice versa.

One of the first examples in the automotive industry was sharing GPS information to understand traffic. Next was driving behavior, a type of vehicle use of location information for *Pay As You Drive* (PAYD) insurance. In the future, we will see information shared for all types of scenarios including weather, road conditions, other vehicle's conditions, commerce, advertising and much more. We developed the figure (see Fig. 1.24) to show the levels and maturity of self-integrating in more detail, followed by the description of each of the levels.

Level 0:   Originally, when vehicles were built, very little to no integration existed. Systems were encapsulated and had little integration with the outside world. An example would be the original analog radios built into vehicles and receiving only a broadcast signal.

Level 1:   When the connected vehicle was first introduced, the basic integration was to a call center where information started to be shared for safety reasons. Other examples included a voice call for information sharing

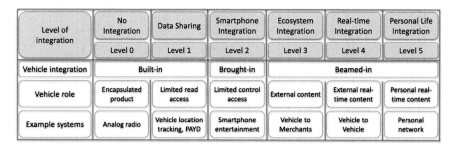

Level of integration	No Integration	Data Sharing	Smartphone Integration	Ecosystem Integration	Real-time Integration	Personal Life Integration
	Level 0	Level 1	Level 2	Level 3	Level 4	Level 5
Vehicle integration	Built-in		Brought-in	Beamed-in		
Vehicle role	Encapsulated product	Limited read access	Limited control access	External content	External real-time content	Personal real-time content
Example systems	Analog radio	Vehicle location tracking, PAYD	Smartphone entertainment	Vehicle to Merchants	Vehicle to Vehicle	Personal network

Fig. 1.24   Levels of the self-integrating foundational theme of a cognitive vehicle

	and PAYD insurance where location and speed information was used for calculating our insurance rates.
Level 2:	As the smartphone industry grew, the ability to integrate or bring the smartphone into the vehicle grew as well. The original scenarios focused on entertainment where the customer had the ability to play their own songs, listen to online radio, or talk on the phone.
Level 3:	As wireless and the cost to integrate wireless into the vehicle started to come down, more scenarios emerged where the vehicle was integrated into the cloud, or data was beamed into the vehicle. Examples include dynamic traffic and weather info, with the next generation including more merchant ecosystem integration where the customer can purchase fuel or order coffee from their vehicle.
Level 4:	*Vehicle-to-vehicle* (V2V) and *vehicle-to-infrastructure* (V2I) are the driving forces behind the real-time integration of the vehicle. The cloud is also a driving force, where open platforms can share information from vehicles in real time to improve safety functions. Examples include vehicle sharing their wiper and temperature information to help weather companies forecast and track the weather.
Level 5:	What does the future hold—how much of our personal information will we allow to be shared? Today and in future, more and more information in our personal network will be integrated into and shared in our private and public ecosystems.

The second enabler for taking care of its occupants is

self-configuring to personalize and customize the environment.

Made famous by Apple, the focus on customer experience is now making its way into the automotive industry. The simple beginning of Driver 1 and Driver 2: seat settings, climate settings, radio and music preferences will explode into a new

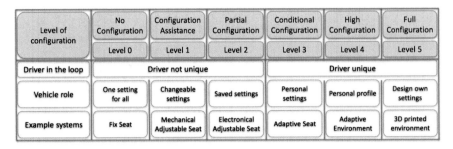

Level of configuration	No Configuration	Configuration Assistance	Partial Configuration	Conditional Configuration	High Configuration	Full Configuration
	Level 0	Level 1	Level 2	Level 3	Level 4	Level 5
Driver in the loop	Driver not unique			Driver unique		
Vehicle role	One setting for all	Changeable settings	Saved settings	Personal settings	Personal profile	Design own settings
Example systems	Fix Seat	Mechanical Adjustable Seat	Electronical Adjustable Seat	Adaptive Seat	Adaptive Environment	3D printed environment

Fig. 1.25 Levels of the self-configuring foundational theme of a cognitive vehicle

type of driving experience. The concept of moving our persona between vehicles, financial transactions from the vehicle to merchants and health monitoring will all be part of a new personalized experience. The vehicle itself can also be configured. What is now an over-the-air (OTA) update will turn into features and functions being downloaded or a vehicle changing from our personal vehicle to a ridesharing vehicle. We developed the figure, (see Fig. 1.25) to show the levels and maturity of self-configuring in more detail, followed by the description of each of the levels.

Level 0:   Henry Ford once said, "You can have any color as long as it is black" [3]. That is a good definition for no configuration options. Maybe a more recent example that is easier to relate to are seats that could not be adjusted.

Level 1:   The configuration assistant or more simply said, the manual seat adjustment. Not many features existed early on that allowed us to change the configuration or setting and make it unique to the driver. Another example were adjustable mirrors.

Level 2:   Who remembers when we pulled the button out on the radio and pushed it back in to save the radio station? This is just one of the first examples of OEMs' thinking and understanding configurations and personal preferences. The seat and climate settings soon followed, but now we could actually save a setting for the next time we returned to the car.

Level 3:   Now we start to transition from being not very unique to a specific driver type configuration. Personal keys were introduced, better known as Driver 1 and Driver 2, but with regard to the radio, entertainment, communication, seats, mirrors, and even driving style, we could now set up the vehicle more to our personal style.

Level 4:   Now imagine more than Driver 1 and Driver 2, but a system that knows exactly who we are and keeps a personal profile, not so far off from what Facebook is already doing. These concepts do not exist in vehicles today, but they are not far off and will only expand with more and more information in the future. As more digital and software capabilities enter the vehicle, being adaptable to the driver is easier to do and will be more of a desired function.

Level 5:   Today we see 3D printing being used in many industries for parts manufacturing. But now imagine going to your dealership and printing out your seats or dashboard configuration, on demand and part of the vehicle purchasing experience. This is a true form of designing for one's own settings and preferences.

But the cognitive vehicle needs also to take care of itself. An elementary foundation is

self-learning to optimize performance for occupants and the environment intelligently.

A new scenario to the automotive industry and other industries is the ability to move beyond collecting and analyzing data with traditional analytics to developing online learning from the experience data (see Fig. 1.14). This can be done for both the vehicle and occupants, and leads to scenarios where the CPA can advise instead of just respond to information and requests. With the improvements in deep learning (see Sect. 1.2.4) we are seeing today, this will be an important capability for providing a real personalized experience in the vehicle. As the vehicle learns more about the driver and the occupants, it will be able to expand its advice to other mobility services options. We developed the figure (see Fig. 1.26) to show the levels and maturity of self-learning in more detail, followed by the description of each of the levels.

Level 0:   Citizen Band Radio (CB Radio) was first introduced in the commercial vehicle industry and was used by drivers for ad-hoc communication

Level of learning	No Learning	Data Sharing	Smartphone Learning	Narrow AI	General AI	Superior intelligence
	Level 0	Level 1	Level 2	Level 3	Level 4	Level 5
Vehicle integration	Built-in		Brought-in	Beamed-in		
Driver role	Unstructured conversation	Speech Synthesis	Basic questions	Provide domain context	Provide personalized context	No human needed
Analytics Level	None	Foundation	Descriptive	Predictive and Prescriptive	AI	Beyond Human Understanding
Example systems	CB radio	TTS	Siri	Nuance	CPA	"Deep Thought"

Fig. 1.26 Levels of the self-learning foundational theme of a cognitive vehicle

while driving across the country. This is a good example of unstructured communication where the learning or understanding is the responsibility of the driver, not the radio itself.

Level 1:   The first systems introduced into the vehicle had voice connections to a call center, either we talked to an agent or an interactive voice response (IVR) system, which used basic voice recognition but were mostly driven by TTS. Later even turn-by-turn directions were sent to the vehicle and TTS was used to guide us to our destination.

Level 2:   Now the smartphone is integrated or brought into the vehicle and along with that comes its most popular applications, in the case of Apple, Siri. Using more advanced voice recognition technology, Siri moved beyond basic commands to answering basic questions. Siri could also learn or understand more of our information on our smartphone, from our calendar to our address book.

Level 3:   The cloud combined with voice recognition could now deliver a better natural language experience combined with access to more domain information. Nuance, a leader in embedded and cloud voice technologies, developed specific domain knowledge about the vehicle and driving. It is similar to Siri, but integrated into the vehicle and can be considered what we defined earlier as Narrow AI.

Level 4:   We can already see today with Alexa and Siri the core capabilities of a CPA, but what does the future hold? Sure Siri knows a lot about us, but it is usually in a single domain, like our calendar—recommending bringing an umbrella to go to our meeting on Friday, because it is supposed to rain. More and more, Siri and the other CPAs will start sharing context and building a more personalized experience across our domains, a path towards real General AI solutions.

Level 5:   Super intelligence, will human thinking and learning become extinct? We already talked about the fictional computer "Deep Thought", which is very capable of such thinking. However, looking at it realistically, could CPAs become intelligent to the point where our human communication is no longer required?

The other foundation is

self-healing to analyze and predict services and maintenance of the vehicle.

A long-time scenario in the automotive industry plus in all aspects of the IoT, is collecting machine information to help with service, maintenance and ultimately improve on predicting failure. This scenario has one of the highest returns on investments and is one of the most mature in all industries. Even before automotive the other industries were monitoring remote equipment and collecting data on usage

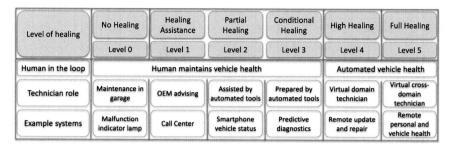

Level of healing	No Healing	Healing Assistance	Partial Healing	Conditional Healing	High Healing	Full Healing
	Level 0	Level 1	Level 2	Level 3	Level 4	Level 5
Human in the loop	Human maintains vehicle health				Automated vehicle health	
Technician role	Maintenance in garage	OEM advising	Assisted by automated tools	Prepared by automated tools	Virtual domain technician	Virtual cross-domain technician
Example systems	Malfunction indicator lamp	Call Center	Smartphone vehicle status	Predictive diagnostics	Remote update and repair	Remote personal and vehicle health

Fig. 1.27 Levels of the self-healing foundational theme of a cognitive vehicle

to improve maintenance. The improvements in machine learning techniques and the ability and reduced effort to implement them more effectively are an important scenario moving forward. Vehicles will be able to fix and optimize themselves based on certain events or situations without human intervention. We developed the figure (see Fig. 1.27) to show the levels and maturity of self-healing in more detail, followed by the description of each of the levels.

Level 0: A light goes off. I'm not sure what to do. Take it to the dealer and let them figure it out. Before that, the vehicle pretty much had to stop working before people took it to the dealer. The vehicle could tell us about a problem, but it could not heal itself in any way. That was the responsibility of the technician at the dealer.

Level 1: One of the most convenient features first introduced into vehicles was the check engine light coming on in our vehicle and we could talk to someone over the phone and determine how serious the problem was. Do we need to pull over or could the problem wait until our next scheduled maintenance? This was a huge timesaver. The technician still had to fix all the problems at the dealership, but with the help of the OEM and the connected vehicle it didn't always have to ruin our day.

Level 2: As the solutions matured, more data was collected from the vehicle and the OEM could potentially learn more about our problems. With smartphones and the web, vehicle status could be viewed in real time or a monthly health report was sent out so we could be kept up-to-date on the health of our vehicle. The dealer and technician were still in the loop, but the customer was more informed.

Level 3: Along with collecting more data and the evolution of analytics tools, OEM could now do more than just send a health report, but it also had the ability to predict when a problem might occur. More common in commercial telematics, but it is still a very useful function in the consumer space and with helping the OEMs with warranty costs.

Level 4: In the future, the concept of a virtual technician comes into play as we transition from human-assisted to automated healing. Even today, some software problems in the vehicle can be handled with updates and no interaction with the dealer. As the vehicle transitions from a mechanical system to more of a software system, this type of healing and repair will become more commonplace.

Level 5: Once our vehicle is healthy and automatically monitored, our health and the health of our devices and environments around us is the next logical step. Parts of this exist in many industries, but putting this all together in a comprehensive health and monitoring solution will be a goal for many industry OEMs.

The general opinion is that a

self-driving car (also known as autonomous car) is one that is capable of sensing its environment and navigating without human input.

A very popular scenario in the press over the last year, with capabilities being announced by many OEMs and Tesla Motors no longer being seen as the only leader.[20] It is important to separate the hype from reality, and most discussions in the press are about Level 2 or 3, not Level 5 following the Society of Automotive Engineers (SAE) standard for automated driving classifications (see Fig. 1.28)[21]:

Level 0: Automated system has no vehicle control, but may issue warnings.
Level 1: Driver must be ready to take control at any time. Automated system may include features such as adaptive cruise control, parking assistance with automated steering, or lane keeping assistance in any combination.

Level of driving automation	No Automation	Driver Assistance	Partial Automation	Conditional Automation	High Automation	Full Automation
	Level 0	Level 1	Level 2	Level 3	Level 4	Level 5
Driver in the loop	Human driver monitors driving environment			AD system monitors driving environment		
Driver role	Must drive	Can drive with assistant	Must monitor driving	Must be prepared to intervene	Driver may sleep	No driver needed
Example systems	Lane departure warning	Lane keeping assistant	Traffic jam assistant	Automated parking	Highway driving pilot	Automated taxi

Fig. 1.28 Levels of automated driving pursuant to the SAE international standard J3016

[20]"Detroit Is Stomping Silicon Valley in the Self-Driving Car Race" https://www.wired.com/2017/04/detroit-stomping-silicon-valley-self-driving-car-race/. Accessed: July 26, 2017.

[21]"Automated Driving: Levels of driving automation are defined in new SAE international standard J3016" https://www.sae.org/misc/pdfs/automated_driving.pdf. Accessed: July 26, 2017.

Level 2:   The driver is obliged to detect objects and events and respond if the automated system fails to respond properly. The automated system executes accelerating, braking, and steering, for example, as a traffic jam assistant. The automated system can deactivate immediately upon takeover by the driver.

Level 3:   Within known limited environments (such as highways), drivers can safely turn their attention away from driving tasks by reading a book, texting or web surfing, but must still be prepared to take control when needed. Automated systems may include features such as Traffic Jam Pilot or automated parking.

Level 4:   The automated system can control the vehicle in all but a few environments such as severe weather. The driver must enable the automated system only when it is safe to do so. When enabled, driver attention is not required and driver may even sleep. Such automated systems may be used on a closed campus as a driverless shuttle or for valet parking in garages.

Level 5:   Other than setting the destination and starting the system, no human intervention is required. The automated system can drive to any location where it is legal to drive and make its own decision, such as an automated taxi.

Even with the current focus on Level 2 or 3, vehicles will become highly automated with some areas of limited autonomous function in controlled environments.

We should notice that the cognitive vehicle definition is not exactly what we have listed as a "robot on its own" in Fig. 1.7. It has self-driving capabilities like a "robot taxi", but a personalized mobility experience is much more than that, according to what we have seen in the six foundations (see Fig. 1.23). We will dive deeper into this discussion in the next chapter.

Furthermore, a cognitive vehicle cannot be alone, it has to work with others. One foundation is

self-socializing is the vehicle's social networks to assist others and utilizing the vehicle for ancillary tasks.

Beyond people socializing from vehicles, sharing location or having their email or Short Message Service (SMS) message read, or talking hands-free on their phones, the vehicle itself will begin to connect to the greater IoT and socialize with devices from other industries. In the future, sharing information with other vehicles for safety reasons and with cities for traffic and accident reporting will create an entire new ecosystem of vehicle data in the cloud and between vehicles. For the sake of this book and because of the high costs and the slow speed of adopting standards related to V2V and V2I, see [10] for more details. We will refer to this area

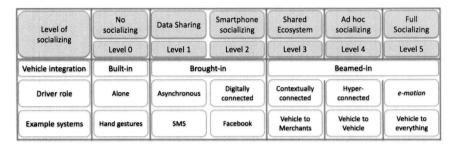

Level of socializing	No socializing	Data Sharing	Smartphone socializing	Shared Ecosystem	Ad hoc socializing	Full Socializing
	Level 0	Level 1	Level 2	Level 3	Level 4	Level 5
Vehicle integration	Built-in	Brought-in		Beamed-in		
Driver role	Alone	Asynchronous	Digitally connected	Contextually connected	Hyper-connected	*e-motion*
Example systems	Hand gestures	SMS	Facebook	Vehicle to Merchants	Vehicle to Vehicle	Vehicle to everything

Fig. 1.29 Levels of the self-socializing foundational theme of a cognitive vehicle

only where the data are shared in the cloud. We developed the figure (see Fig. 1.29) to show the levels and maturity of self-socializing in more detail, followed by the description of each of the levels.

Level 0:   How did people socialize before smartphones? Most of us cannot remember, however simple meetings and conversations were common before everyone's head became buried in their smartphones. While driving, the most common form of communication was gestures or signaling with our horn or lights. Unfortunately it was more common to react negatively than positively, but we will not be getting into road rage.

Level 1:   The first cellular phones and basic smartphones focused on a form of digital communication called SMS, or texting. It is still common today, and to the point where some states have issued texting-while-driving laws to stop people from using their smartphones while driving.

Level 2:   It wasn't long after SMS that pictures and video sharing became popular, and social media companies like Facebook took it to the next level and created a fully digital sharing experience online. Our digital persona was born and now people could connect and socialize in real time and even do video conferencing from halfway around the world.

Level 3:   Now that we have a digital persona and we could connect with our friends, what about merchants and brands that were important to us? Facebook started to implement ways we could share or use our digital persona even to log in to other merchants' sites creating a digital ecosystem around our persona as well.

Level 4:   In the future, vehicles will become part of social networks as well, unlike Facebook but using similar technologies. We already talked about V2V and V2I as examples where vehicles share data in real time and in an ad-hoc fashion.

Level 5:   *e-motion*, we introduced this earlier, but without a definition. Simply, it is the e-business way of talking about the new emotional connection between brands and customers that is moving beyond physical and becoming more digital with a focus on the customer experience and creativity with a human touch.

1.7   e-Motion

We introduced *e-motion* as the highest level of the self-socializing fundamental theme of a cognitive vehicle. Put simply, it is the e-business way of talking about the new emotional connection between brands and customers that is moving beyond physical and becoming more digital with a focus on the customer experience and creativity with a human touch.

Basic emotions are positive or negative, and intense or mild. But e-motion goes deeper. In the last decades [12], many researchers have agreed that there are six universal emotions, which are distinctly associated with facial expressions for all human beings (see Fig. 1.30).

- Anger
- Disgust
- Fear
- Happiness
- Sadness
- Surprise

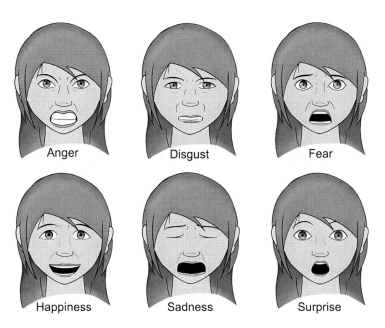

Fig. 1.30 Six universal emotions which are distinctly associated with facial expressions for all human beings

It is interesting to note that four out of the six are negative emotions. This e-motion leads us to the beginning of the introduction (see page 2) when we asked the question:

►   Do we LOVE our cars?

In the sense of the six universal emotions, love is happiness. But love is also trust, which is not covered by the facial expressions. Now e-motion is becoming more complex. Today, more than 154 different emotions and feelings have been defined, and there is a complex array of overlapping words in our languages to describe them. In [18] Robert Plutchik said that the feeling state is part of a process involving both cognition and behavior and containing several feedback loops. He created the wheel of emotions (see Fig. 1.31), from which we concluded that the human feeling of love is emotional joy (happiness) and trust. Plutchik suggested eight primary bipolar emotions depicted in this wheel: joy versus sadness; anger

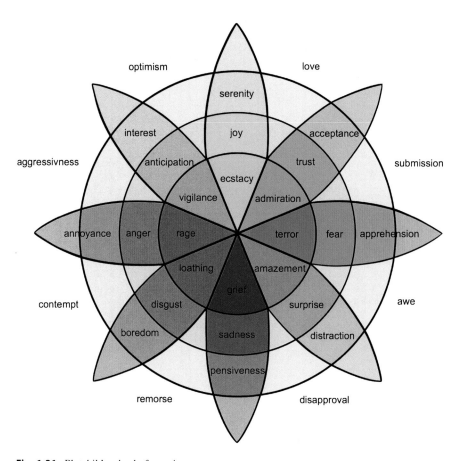

Fig. 1.31 Plutchik's wheel of emotions

versus fear; trust versus disgust; and surprise versus anticipation. Additionally, his circumflex model makes connections between the idea of an emotion circle and a color wheel. Like colors, primary emotions can be expressed at different intensities and can mix with one another to form different emotions.

Now this is the challenge the CPA is facing to achieve 'My Cognitive autoMO-BILE Life.' We need to design the CPA as the "command center" of our e-motion. It assesses stimuli from traveling in the outside world, deciding whether to avoid or approach, in addition to not only identifying tone or facial expressions.

That will lead us to the generation gap between a traditionalist and a digital native or even a hyper-connected, where a traditionalist says

► *It is a computer and that is it. It doesn't love you.*

but the kid's reaction is

► *"Why is it right to love the teddy bear and not to love my cognitive personal assistant?"*

1.8 Structure of the Book

This book deals with defining a cognitive vehicle and a cognitive life in terms of their potential to disrupt the automotive industry. It is divided into four chapters:

1. The introduction starts out with the fundamental question around these new CPAs in the market and how far we will let them become part of our lives. This creates a problem for traditional OEMs as the industry moves towards a cognitive vehicle and on to a cognitive life because their need to transition from the physical world to the personal digital world. The next sections define some key capabilities around analytics, AI, customer experience and microservices, separating the reality from the hype. We then define the cognitive vehicle based on self-enabling themes where e-motion is defined as the personal way in which customers and brands connect.
2. The platform for a cognitive vehicle life is a journey through history and into the future as we look at the services domains and capabilities that technically built the connected vehicle. We use the self-enabling themes to measure the industries progress through time and detail the maturity levels for each theme. The automotive industry has learned much over the years and we publish these lessons to help other industries currently undergoing a product-to-services transformation. We conclude the chapter with some interesting analysis on how a cognitive life is now more of a life-centric model than a vehicle-centric model. Smartphone brands are better positioned than traditional OEMs to support the digital platforms of the future and are strategically moving into more touchpoints with customers.

3. The value chain and business models move away from technology and focus on the investments needed to drive both OEMs and customer value. We explore subscription models, freemiums, revenue sharing, on-demand and the newly developed model for a cognitive life. We introduce a framework that aligns with the self-enabling themes so OEMs can set goals and track progress for both internal and external value. We show how some OEMs are making progress towards change, but analyze if their changes are quick enough to compete with the likes of Apple and Google.

4. The challenges in my cognitive autoMOBILE life dig deeper into the current challenges in the CPA market and where it is clearly headed to support a customer-centric life. We understand that trust and privacy will be primary concerns for customers moving forward and we look at blockchain technology as an option to bring the type of transparency, integrity and security needed to support the future vehicle digital platform. Autonomous driving will struggle in the short term, but with the mobility industry evolving from ridesharing to ownership sharing, this industry may see some of the first implementations. We wrap it up with e-motion, and what everyone needs to think about as the industry changes around us.

1.9   Who Should Read This Book?

The scope of this book is primarily aimed at executives in the automotive industry or new industries related to the IoT, who are developing an implementation plan for new or improved business possibilities, and who want to understand the market opportunities offered by designing data-driven business models. But building a strategy, enhancing disruptive business models, and developing the company's culture is a big task in the automotive industry that cannot be done by one role or even as a part-time job. Many car manufactures have recently introduced the position of *Chief Digital Officer* to their executive management, whose professional profile shows a deep understanding of technology and its interaction with humans, culture and strategy. This book addresses this role in particular but does not assume any knowledge about artificial intelligence and data-driven techniques. Its goal is to provide concepts and tools needed to design systems capable of learning from autoMOBILE data in this new and rapidly shaping mobility industry.

1.10   Why Did We Write This Book?

We both started thinking about bringing cognitive capabilities into the autoMOBILE architecture in May of 2012, when we were working to help AUTOmobile manufacturers understand the values of digital connectivity and personalization through data and AI capabilities. The industry and press didn't help by creating more hype and false definitions around AI and how systems learned, or in reality how systems are trained. It resulted in a long-term exercise to disrupt the day-to-day thinking of customers and people in general, and understand this new and developing area. At the same time, the IoT was also getting a lot of hype in the press and for a solution that had actually been around for probably 20 years. They made it sound as if had just been invented. Looking at the automotive industry and its most recent transitions into cognitive vehicles, the lessons learned for other industries and companies just entering the field of IoT solutions seemed valuable to share, since AUTOmobiles have been connected since 1996. Finally, we both worked together with many global automotive and electronics clients on our visions about cognitive autoMOBILE life during our international assignments in Japan from 2015 to 2017. In 2016, we decided to write this book, which would help navigate the wide range of topics related to the e-motion of the cognitive vehicle and the impact on our lives, which goes far beyond the values and capabilities of the autonomous driving being discussed today.

References

1. Brown T (2009) Change by design: how design thinking transforms organizations and inspires innovation. HarperCollins, New York
2. Deng L, Dong Y (2014) Deep learning: methods and applications. Found Trends Signal Process 7(3–4):197–387
3. Ford H, Crowther S (1922) My life and work. Garden City, New York
4. Hawking S (1988) A briefer history of time. Bantam Books, New York
5. Hinton G, Osindero S, Teh Y (2006) A fast learning algorithm for deep belief nets. Neural Comput 18:1527–1554
6. IBM Institute for Business Value (2013) Digital reinvention: preparing for a very different tomorrow. IBM Corporation, Somers
7. IBM Institute for Business Value (2015) Automotive 2025: industry without borders. IBM Corporation, Somers
8. IBM Institute for Business Value (2016) A new relationship – people and cars. IBM Corporation, Somers
9. IBM Institute for Business Value (2016) Who's leading the cognitive pack in digital operations? IBM Corporation, Somers
10. Jurgen RK (ed) (2012) V2V/V2I communications for improved road safety and efficiency. SAE International, Warrendale
11. Masse M (2011) REST API design rulebook. O'Reilly, Sebastopol, CA
12. Matsumoto D (1992) More evidence for the universality of a contempt expression. Motiv Emot (Springer Netherlands) 16(4):363–368
13. Minsky M (1961) Steps toward artificial intelligence. In: Proceedings of the IRE. https://doi.org/10.1109/JRPROC.1961.287775
14. Newman S (2015) Building microservices – designing fine-grained systems. O'Reilly, Sebastopol
15. Nickels M (1986) Eliza. Markt and Technik, Sonderheft 4/1986, 49–57
16. Norman DA (2004) Emotional design: why we love (or hate) everyday things. Basic Books, New York
17. Pfister B, Kaufmann T (2017) Sprachverarbeitung: Grundlagen und Methoden der Sprachsynthese und Spracherkennung, 2. Aufl. Springer, Berlin
18. Plutchik R (2001) The nature of emotions. Am Sci 89(4):344–350
19. Russell SJ, Norvig P (2009) Artificial intelligence: a modern approach, 3rd edn. Pearson, Essex
20. Salus PH (1994) A quarter century of UNIX. Addison-Wesley, New York

21. Samuel AL (1959) Some studies in machine learning using the game of checkers. IBM J 3(3):535–554
22. Schulte WR, Natis YV (1996) "Service Oriented" architectures, part 1. Gartner Research, Stamford, CT
23. Shevat A (2017) Designing bots. O'Reilly Media, Sebastopol
24. Sternberg RJ, Pretz JE (eds) (2005) Cognition and intelligence: identifying the mechanisms of the mind. Cambridge University Press, New York
25. Turing AM (1950) Computing machinery and intelligence. Mind LIX(236):433–460
26. Wedeniwski S (2015) The mobility revolution in the automotive industry: how not to miss the digital turnpike. Springer, Heidelberg
27. Winterhoff M et al (2015) Think act: automotive 4.0. Roland Berger Strategy Consultants, Bloomfield Hills

Platform for a Cognitive Vehicle Life

2

As we start this technical section and go back in history to talk about how a non-connected vehicle has turned into a *cognitive vehicle*, we will start with the core service domains and capabilities, develop them to where we are today, and end with what we may see in the future. We have seen many different definitions and architectures for the industry. What we will try to do is build the capabilities, not to a perfect timeline for any particular OEM, but to highlight the key components and how they affected the growth of each generation. Our architectural focus will not be on privacy and other non-functional requirements like security, scalability, reliability, etc., but more on the types of customer experience, services and the capabilities to deliver them. One more comment about privacy and security: The industry has certainly given these issues more attention now in this digital era than it did in past generations. But starting with Generation 3, this focus is needed more as they enter the fully digital world. At the end, we will tie those to how other industries can build similar solutions and join a connected ecosystem.

We structured the evolution of these core service domains and capabilities into the following five generations:

Generation 0:   This is how it all began before the vehicle became digitally connected.

Generation 1:   The first wireless connectivity focusing on the initial area of services for connected vehicles called safety and security.

Generation 2:   The first digital integration focused on navigation and infotainment and their introduction and integration into the vehicle. Besides passenger vehicles, the fleet or commercial businesses adapted the connectivity with a stronger focus on revenue and profit.

Generation 3:   Moving from digital to intelligence, this generation introduces the cognitive vehicle as it is today and as it may look like in 2022 when learning and a personalized experience have been developed.

Generation 4:   Cognitive life where connectivity and a persona follows our e-motion around.

© Springer-Verlag GmbH Germany 2017
S. Wedeniwski, S. Perun, *My Cognitive autoMOBILE Life*,
https://doi.org/10.1007/978-3-662-54677-2_2

For each generation, we will summarize a solution architecture structured into three layers, the vehicle, the Internet of Things (IoT) and the cloud, (see Fig. 2.1). The middle layer,

> the *Internet of Things* (IoT), is the interlinking of connected physical devices, especially in our case the vehicles in the first layer, and other items embedded with electronics, software, sensors, actuators, and network connectivity which enable these objects to collect and exchange data.

The major difference of the IoT from the older term *machine-to-machine* (M2M) is the cloud, (see Fig. 2.2). M2M refers to direct communication between devices using any communication channel, including wired and wireless, what is V2V in automotive, for example. Therefore, thanks to the boundless capabilities of the cloud, the IoT connectivity of devices and services is more advanced than the limited direct M2M communications like exchanging data via Bluetooth.

The mobility cloud is a separate layer from the IoT in the solution architecture to differentiate the type of service domains. The IoT layer has cloud computing capabilities that provide shared computer processing resources and data. But it's

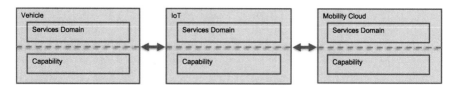

Fig. 2.1 Solution architecture pattern structured in the three layers—vehicle, IoT, and cloud

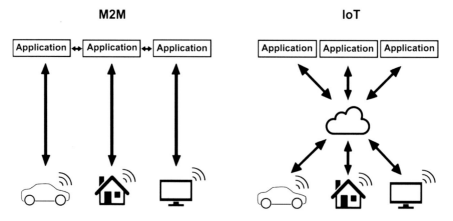

Fig. 2.2 The IoT leverages the services in the cloud for interlinking connected devices, rather than using direct communication—wired and wireless—between devices in M2M

more managing the connected physical vehicle and deployable services. Instead the Mobility Cloud is treating the vehicle more virtually and defines the services needed for the personalized mobility experience.

We will break down each layer of the solution architecture into domains of services and capabilities. The domains of services are assessed based on use cases from the user's perspective, and the capabilities are structured by technologies or business processes required to deliver the solution. Furthermore, we will discuss how the solution can be applied to other industries.

Our focus for evolving the different solution architecture generations is to develop the cognitive vehicle and how it can make our life so much better, also introduced as the autoMOBILE life. Especially in Sect. 1.6, we defined the cognitive vehicle with the foundation of the six self-enabling themes, which we will use to assess the maturity of each generation towards 'My Cognitive autoMOBILE Life.'

- Self-integrating,
- Self-configuring,
- Self-learning,
- Self-healing,
- Self-driving, and
- Self-socializing.

As we go through the history of the connected vehicle, think about the industry and the products that are being manufactured today, think about how it may change, think about how a company may change because of it. We offer examples of each generation and discuss other products and industries in hope of helping to create a broader understanding for transforming better and starting to learn *e-motion*. Remember some keywords that drive customer experience *e-motion* that still are factors in the digital and connected world today, convenience and safety are two to think about and watch transform.

2.1 Generation 0: The Non-connected Vehicle

► Do you remember when your vehicle wasn't connected?

That is probably not that hard since most vehicles aren't really connected today, but connected by having a smartphone, that might be a bit different.

► But do you remember your life before the smartphone?

In 1990s we developed systems to support processes in manufacturing to standardize the interfaces between various production machines, at that time called computer integrated manufacturing [9], now we are calling it Industry 4.0[1] or the IoT for Manufacturing. After the first systems were developed, we got pagers to be notified if we had any problems, especially when they occurred on the third shift. That is great connectivity, until the first night when we are paged at 3 a.m. Nowadays, we have a smartphone, which is basically connected all the time. Messages in the morning, email in the evening. Certainly it's more work-life integration than balance, without any separation between vacations and work since the device is always connected to both work and personal matters, in either way, hyper-connected might be the word, far beyond the pager from the job experiences in 1990s.

What was the vehicle before connectivity, mobile entertainment maybe? The analog radio was the first type of entertainment added to vehicles. Then came 8-track tapes, cassette tapes and compact discs. OEMs thus did understand customer experience, bringing content into the vehicle, and entertaining people. But also at that time, most of the customer experience was driven by physical things. As we first discussed in the introduction, engine size, color, style, all physical elements were the primary factors driving customer experience and buying decisions.

Beyond entertainment, safety and convenience became popular items as well. As more and more embedded electronics entered the vehicle, anti-lock braking system, power seats, power windows, and even electronic fuel injection and transmission controls all became popular items to drive the buying decisions.

2.1.1 Solution Architecture

As established, the vehicle, the IoT and the cloud make up the three entities required to build a connected vehicle solution, or in the case the non-connected vehicle. The architecture model (see Fig. 2.3) reflects how early devices were integrated into the vehicle for basic communication and entertainment. Starting out with the vehicle, it focuses on the capabilities delivered by an analog radio or a CB radio, basically the ability to send or receive voice communication. The IoT was simply the carrier or radio transmission, and the mobility cloud being the stations with the broadcasters and the content.

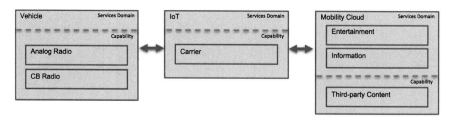

Fig. 2.3 The solution architecture of service domains and capabilities for Generation 0—the non-connected vehicle

---

[1]"Platform Industry 4.0" http://www.plattform-i40.de. Accessed: July 26, 2017.

2.1.2 Vehicle

In the vehicle layer, we didn't define any service domains but two capabilities:

1. Analog radio to receive broadcasted information and entertainment in the vehicle,
2. and CB radio for communications.

Analog Radio
Analog radio was first used in homes and later made compact enough to fit into vehicles. It received radio broadcast signals from regional stations with content that included local activities, news, weather, and music for entertainment.

CB Radio
A bi-directional analog radio used mostly in the commercial vehicle industry for localized communications and in a way entertainment for the long-haul industry as drivers traveled across the country. On the consumer side, it was the foundation for the first type of social networks as well, when friends traveled in groups or around their communities, they could easily communicate with this low-cost and license-free technology.

2.1.3 Internet of Things

There was no IoT or even an Internet at the time of this generation, but we want to keep the solution architecture pattern common across the different generations and actually the defined capability 'carrier' fits into this layer from today perspective.

Carrier
The first carriers were local radio stations that broadcast the radio waves received by analog radios. Two types of signals existed, amplitude modulation and frequency modulation, each had a frequency range that was aligned to the radio, and consumers could tune in a specific station and listen to the content.

2.1.4 Mobility Cloud

In the mobility layer, we defined two service domains that are consumed by the vehicle layer:

1. Entertainment was primarily audio music to limit driver distraction,
2. Information was mostly generic and in some cases developed for local content, but never personalized in this generation.

Both service domains were heavily dependent on the capabilities of the third-party content.

Entertainment

The first type of entertainment in the vehicle was music, which is still popular today, the only difference today being the type of technology used. Many stations existed which delivered different types of music genres, for example, blues, country, jazz, etc. Talk shows also became popular as a form of entertainment in some cases or information in others.

Information

The first type of information in the vehicle was news, sports and weather, followed by traffic. Advertising was the business model and the content was interspersed with advertisements for local business or activities within the community.

Third-Party Content

The radio station was the broadcaster, the entertainment and information all came from third parties. Staff or reporters would travel around or make calls to their providers, wrote their segments and then it was broadcast over the air.

2.1.5   How Are These Services Used by Customers?

It is amazing that the solutions and technologies introduced so long ago are still popular today. Everything is going digital and some content even comes at a cost, but the fundamental idea of receiving entertainment and communicating in the vehicle remains the same and what a better way to pass the time while driving.

2.1.6   How Far Is It from 'My Cognitive autoMOBILE Life'?

We are assessing the Generation 0 platform based on the foundation of the six self-enabling themes (see Sect. 1.6) and illustrate the score in Fig. 2.4:

Self-integrating	0, for no digital integration and information sharing.
Self-configuring	1, for manually changeable settings to customize some AUTO-mobile components.
Self-learning	0, for no intelligent optimization at all.
Self-healing	0, for only straight human-based analysis and maintenance in a garage.
Self-driving	0, for no automation at all.
Self-socializing	0, for no social networks for vehicles at all.

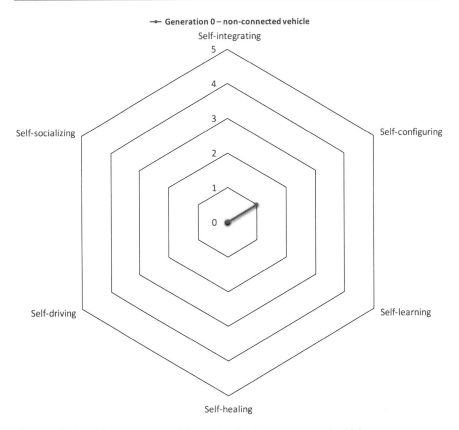

Fig. 2.4 Self-enabling assessment of Generation 0—the non-connected vehicle

2.1.7   How Relevant Is It Across Industries?

Even early on, OEMs realized the value of entertainment and communication. As we read through the generations, we will identify the growth and changes of these areas and their effect on the industry. Advertising was the business model for radio stations and not much has changed there either, with most of the Internet-based companies still using this model today.

2.2   Generation 1: Connected Vehicle

Basically, the connected vehicle is equipped with Internet access. But actually only the combination of vehicle, wireless technology and cloud all together form the connected vehicle solution. Otherwise, it would be just an Internet web-browsing access feature in the car. There are different claims as to who built the first connected vehicle. For example, BMW is claiming the first connected mobility solution through radar technology used for monitoring braking distances, utilized in their concept sports car E25 Turbo in 1972. Actually, the core technology "telematics" was coined in the late 1970s as a result of the increasing interconnections between computers and telecommunications media as a combination of the words "*Tele*communication" and "Infor*matics*". But the development of telematics is the result of consumer-oriented electronics like computers, mobile phones and multimedia. At that time, it wasn't really considered for bringing or embedding electronics into the AUTOmobile.

The Internet and telematics began to take center stage in the 1990s, and this is what we will use to define the era of the connected vehicle. At the end, it does not matter who was really first, our focus is much more on what relevant services were provided to lead the connected vehicle solution architecture. In that sense, General Motors was the first OEM to bring the first connected car features to market with OnStar in 1996. The primary focus was driver safety and this beginning started to shape the connected vehicle industry as we know it today. Figure 2.5 is an example of how Safety & Security was implemented as three buttons in a vehicle with most vehicles using a similar layout. The broadening of the telematics systems by using initial *Global Positioning System* (GPS—a global system for determining locations using satellites) navigation devices made it possible to make targeted traffic information available to logistics companies. Until the end of the 1990s, the first implementations of mobile offices or Internet connections were only possible up to a point. The technologies and protocols available at that time, such as *Wireless Application Protocol* (WAP), had slow transfer rates[2] and long response times from the mobile network, which meant that they were not very convenient to use. Furthermore, the display and operating options of mobile phones were very limited at that time. For a more in-depth look at telematics, see the literature [5].

After the success of telematics service provider (TSP) OnStar, many OEMs followed with similar connected emergency and information services with call centers that could be reached at the touch of a button from the vehicle.

Fig. 2.5 Safety & Security services as three-button layout in a vehicle

> Telematics service providers (TSPs) were the first companies other than
> OEMs that set up telematics businesses and sold their services, specifically
> Safety & Security.

At that time, General Motors was the only company that built their own platform,
all other OEMs purchased services through independent TSPs. Actually, different
companies were involved between the vehicle and the actual telematics service:

1. Manufacturers of telematics units in the vehicle
2. Telecommunications network providers
3. TSPs and
4. Providers of information included with the telematics services.

The responsibilities became concentrated in the market, as shown by the fact that
original telecommunications companies such as Ericsson or Verizon have expanded
to also become TSPs. TSP examples in the US market include Verizon Telematics,
Wireless Car, and SiriusXM.

Wireless Car is an example of a TSP that tried to introduce standards and
common APIs for OEMs. In 2010, BMW and Wireless Car in particular have driven
forward the "next-generation telematics patterns" (NGTP) in order to make more
technology-neutral interfaces possible within the infrastructure, and to enable more
flexible implementations of new telematics services, more independently of the TSP.
They have officially added a few more members but broad industry adoption has
not happened. But what is unique about the NGTP is the standard interface at key
boundary points in the architecture. Simple examples include "content providers"
(see Fig. 2.6) where a partner selection for weather will end up with a set of APIs
were development changes and the impact is limited to a change in the partner. The
same holds true for the other telematics services.

The framework of the NGTP is defined by the following four main components
(see Fig. 2.6):

1. The telematics unit is built into the vehicle, and nowadays is often connected to
   the "infotainment" system in via the vehicle's network architecture, which we
   will more deeply discuss in the next generation. It collects data from the vehicle
   itself and receives information from the outside, which it prepares for display.
   Via a wireless network, the telematics unit communicates bi-directionally with a
   "dispatcher".
2. The role of the network layer is to pass messages across network borders,
   between the telematics unit at the "front end" and the downstream telematics
   system at the "back end". The "dispatcher" is the entry and exit channel for both
   end systems, which are connected via local, regional and long-distance networks
   depending on the situation. Independently of the telematics unit, the region in

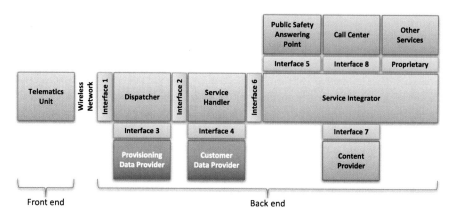

Fig. 2.6   Components (light blue), data (dark blue), interfaces (green) and telematics services (red) from the NGTP 2.0 telematics architecture [6]

which the vehicle is situated, the type of message and the decoupled regulation "provisioning data provider", it finds the "service handler" for the message. Interface 1 serves to decouple the network layer from the telematics unit. As a result, the provider of the network layer can be exchanged, or expanded to new regions. The telematics unit can also vary depending on the vehicle, region and running time in question, and must be independent from the provider or the distributor.

3. Depending on the origin of the telematics unit, the user and any given rights, the "service handler" can add vehicle and user information to the content of the message, or alter it. He makes it possible to interpret manufacturer-specific information in the messages, and to delete specific details from them which manufacturers do not wish to share with TSPs. Interfaces 2, 4 and 6 make it possible to decouple customer data "customer data provider" from the TSP.

4. The "service integrator" brings together telematics services. In the overall system, several integrators can be included at the same time—while acting independently of one another depending on their region. In particular, the "public safety answering point" is a regional emergency call service which is separated especially, using interface 5. The additional dedicated services "call center" and "content provider" for information or support can be made global or regional depending on the user profile, and are also decoupled from the integrator using interfaces 7 and 8. The telematics "other services" are not specified any more precisely, and can be manufacturer-specific proprietary developments for the purpose of differentiation in the market.

For a more detailed description of the NGTP telematics architecture, see the literature [6].

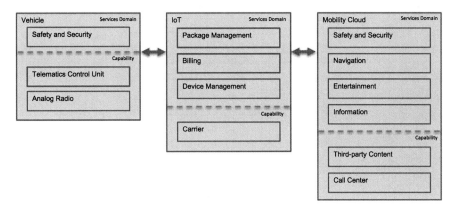

Fig. 2.7 The solution architecture of service domains and capabilities for Generation 1—connected vehicle

2.2.1   Solution Architecture

As established, the vehicle, the IoT and the mobility cloud make up the three layers required to build a connected vehicle solution. The architecture model (see Fig. 2.7) we are introducing has similarities to the NGTP, but we choose not to represent them in this multi-layer style of architecture, instead to more focus on business services and capabilities in a microservices style of architecture for this book, (see Sect. 1.3).

2.2.2   Vehicle

In the vehicle layer, we define the service domain Safety & Security that is also defined in the mobility cloud layer. The services on both layers are very similar but the implementations are different, from the customer perspective the service is seen as a button, from the provider perspective the real service is implemented in the mobility cloud. Hence, for these types of scenarios we will talk about them more in the future, where they are actually implemented.

An important core capability defined by Generation 1 is the telematics control unit to bring about the connected vehicle.

Safety & Security
The domain related to Safety & Security comprises the following services:

Emergency call:	A user in trouble presses a button in the vehicle and is connected to an agent in a call center to assist with the emergency.
Information call:	A user presses a button in the vehicle and is connected to an agent in a call center to assist

	with inquiries or to provide roadside assistance for a flat tire or empty tank.
Stolen vehicle tracking:	A user makes a call to a call center and with a police report asks the agent to track the location of the vehicle.
Automated accident notification:	In the event of a serious accident, the vehicle detects it via sensors and automatically connects to an agent in a call center or even coordinates with a local emergency agency. The agent dispatches emergency responders after talking with the user or even when communication is not possible.
Remote diagnostics:	A user is notified of a problem with the vehicle, typically by a light on the dashboard, and connects to a call center to determine what the problem and its severity is. The check engine light is the most common notification a driver has for an issue with the vehicle. The check engine light is typically triggered by a diagnostic trouble code (DTC), which is a message from onboard computers notifying the driver that something is wrong, see [8] for more details. Not all problems cause a DTC, and it is not always a one-to-one correlation of a problem to a DTC, but it's usually a start.
Remote functions:	A user connects to a call center to unlock the doors of his/her vehicle. Before smartphones, this was the initial way to execute remote functions in a vehicle, later on websites were available for users to do it themselves and to have access to the vehicle's status information.

Most of these services are human-based and provided from the mobility cloud layer. Therefore, consumer-centric APIs are very limited within the vehicle layer for this generation.

Action	URL	Description
GET	/device/{id}/diagnostic	Display diagnostics by lamp ID

Telematics Control Unit

The vehicle's core capabilities come from:

> The Telematics Control Unit (TCU) is an embedded device on board a vehicle that handles the sensing of events and is the external interface for mobile communication.

The TCU contains a CPU, memory, a modem, sensors, an interface to the *Controller Area Network* (CAN bus), and the code for the applications in the service domain. The TCU has access its own sensors like GPS, plus a connection to a bus system interconnecting all the distributed control units in the vehicle called the CAN bus. Most of the other vehicle sensors are connected to other ECUs which communicate to the TCU over the CAN bus. Hence, the TCU does not have direct access to the raw data from most of the vehicle sensors. Only the CAN bus system gives the TCU access to vehicle information, vehicle control and vehicle sensors data. Therefore, the raw data, which is recorded by a sensor, has to go through an analog to digital transformer, a data filter and the local ECU on the CAN bus, before the remaining fraction of data can be read by the central TCU. The CPU runs the applications and manages the interfaces to all the systems and the modems communications. The modem provides the connectivity through the Subscriber Identity Module (SIM) card to the carrier network which we will define in the IoT layer.

The APIs available for a TCU at this time were modeled around the vehicle only and connecting/disconnecting to the IoT layer for sending data or opening a voice call. Basic information is sent VIN and location for an event like an emergency call or an information call. Initially devices could receive controlling commands as well for door unlocking etc. Example operations to connect a vehicle and to send events and receive controlling commands:

Action	URL	Description
PATCH	/vehicle/{id}/connect	Confirm SIM voice or data connection of VIN
POST	/vehicle/{id}/event	Send event of VIN with GPS and data
POST	/vehicle/{id}/service	Instruct vehicle to perform action
POST	/device/{id}/command	Send controlling command to device

2.2.3   Internet of Things

For the IoT layer, the core capability comes from the carrier and wireless network, managing all the connectivity to enable the service domains 'package management,' 'billing,' and 'device management.' The IoT is the center of the business processes and financials required to build the solution, most OEMs rely on TSPs to provide this function.

An early lesson learned about this pattern is to support multiple devices and services related to devices. One industry that has tried to implement standards for connectivity is home automation, the largest organized group is the 'Open Connectivity Foundation',[3] but with the likes of Amazon, Google, and Apple all with their own competing standards and market dominance, this will be a difficult industry to standardize. Even with the lack of standards, a good reference pattern, will allow us to support multiple vendor interactions with our device if we want.

In the IoT layer, we define three service domains:

1. Package management to offer services to the consumers,
2. Billing to manage financial flows for the consumed services, and
3. Device management to assign and identify connected vehicles and users.

The capability of the IoT layer is driven by the carrier, which used to be the telephone company, but is the wireless provider now.

Package Management

In the IoT layer, the service domain 'package management' provides users with the ability to select a package of services, from basic with only one service, for example emergency call, to advanced that includes all the Safety & Security services. Similar to a shopping cart on most commerce websites, it is the ability to select and purchase the features we want. Some features may be in groups, basic packages, others even on an individual basis like only one remote function to unlock the doors. The feasibility of the features depends on the vehicle, make, model, year, and plant of manufacture, which is identified by the globally unique vehicle identification number (VIN).

The APIs for the service domain 'package management' are similar to most commerce websites with a shopping cart. Example operations and select features and packages are:

Action	URL	Description
GET	/features/{id}	List features that are feasible for VIN
GET	/packages/{id}	List packages that are feasible for VIN
GET	/order/cart	Get cart content
POST	/order/cart	Add feature or package to the cart
DELETE	/order/cart	Delete cart item
PATCH	/order/confirm	Confirm order, so the cart contents can be cleared

---

[3]"Open Connectivity Foundation (OCF)" https://openconnectivity.org. Accessed: July 26, 2017.

Billing

In the IoT layer, the service domain 'billing' is primarily used to set up customer and payment information and is closely integrated with the billing system.

Billing includes the core business processes for sending the bill to and collecting from the customer for a package of services. Revenue management is also needed for TSPs as they were contracted by OEMs to implement and provide these initial services. Once a service needs renewal, which is typically triggered by the billing system at the end of a 1 year subscription, for example, the TSP starts the process, reaching out to the customer through email or the call center for renewal of the service. The call center is set up to handle these business processes, along with cancelations, activations of new services, or handling new customer requests. Integration with the billing from a carrier can be required as well, the carrier typically bills for the SIM card, activation, deactivation, data, voice and SMS. Just like a mobile phone bill, these services need to be aligned with the used vehicle services, reconciled and billed accordingly.

These APIs are again like any commerce-related website, once products are selected, bills can be reviewed and paid. Like a mobile phone bill, most people set these up for automatic monthly payment.

1. Example operations regarding customer information, e.g. name, address, credit card:

Action	URL	Description
GET	/account/{id}/address	List addresses by customer ID
POST	/account/{id}/address	Add new address to customer
GET	/account/{id}/address/{id}	List address by ID
PATCH	/account/{id}/address/{id}	Edit address by ID
DELETE	/account/{id}/address/{id}	Delete address by ID
GET	/account/{id}/payment	Get payment method by customer ID
PATCH	/account/{id}/payment	Edit payment method by customer ID

2. Example operations regarding billing confirmation and payment are:

Action	URL	Description
GET	/account/{id}/bill	List bills by customer ID
GET	/account/{id}/bill/{id}	Get bill details by ID
PATCH	/account/{id}/payment	Confirm paid bill if payment succeeded
PATCH	/account/{id}/subscription	Start payment of recurring bills
DELETE	/account/{id}/subscription	End of feature or package subscription
POST	/payment	Make a payment
POST	/payment/{id}	Payment details, e.g. approved/declined
POST	/payment/{id}/void	Void a payment transaction
POST	/payment/{id}/refund	Refund a processed payment

Device Management

Connectivity requires that the modem first be identified and activated on the carrier network, assigned to a specific vehicle, and finally assigned to a specific user. Monitoring is also important, not from the user standpoint, but the device standpoint. Some devices send "heartbeats" to verify to the system they are functioning properly, other types of devices support commands to respond to monitoring type requests. Basic device management can also include configuration or feature updates, turning on a feature, or changing a parameter in a feature, for example to send more or send less data.

Early lessons learned, understanding and designing device management can be difficult. By whom and how a device can be replaced if it does end up faulty is one important aspect. The other is to understand the situations when a device may not be connected, but still is in proper working order. All these types of scenarios need be thought out early in the design, so we don't end up with a large number of devices in the field that need to be replaced, and most likely at a high cost. The APIs available for device management were modeled around the capabilities of the TCU, assigning it to the vehicle and network, and making sure it is healthy and connected via the IoT. Example operations regarding device management are:

Action	URL	Description
GET	/device	List registered devices
POST	/device	Register device by SIM and VIN
DELETE	/device/{id}	Remove device by ID
PATCH	/device/{id}/active	Activate and deactivate device by ID
GET	/device/{id}/config	Get device configuration
PATCH	/device/{id}/config	Update device configuration
GET	/device/{id}/connection	Timestamp of last successful interaction
POST	/device/{id}/heartbeat	Create periodic signal to indicate device operation
DELETE	/device/{id}/heartbeat	Terminate the periodic signal to device
GET	/device/{id}/heartbeat	Ping a device to check normal operation

Carrier

The carrier used to mean a telephone company in the early days of the connected vehicle, but it has evolved into a provider of the wireless network and the connection to the Internet. In the early days of connectivity, the modem used the analog channel, which later became the digital channel most people are familiar with today and refer to as 2G, 3G, 4G, or even 5G. The biggest difference being the amount and speed of data the device could send. Initially on the analog channel, the data were sent as tones over the voice network, just like when calling our favorite airline and entering the frequent flyer number, these are the same type of tones. Later data were transmitted using SMS and then the digital channel we know today.

2.2.4   Mobility Cloud

The mobility cloud's new capability is delivered by a call center, which basically in the first generation was just answering the phone calls from the vehicle. The new service domains are 'Navigation' and Safety & Security. Safety & Security is the back-end of what we described in the vehicle layer.

Navigation
In Generation 1, the navigation service gave directions to desired locations through the call center. The initial part of the call sometime started with voice-based navigation, which didn't always require a human to tell the directions. Example operations regarding navigation are:

Action	URL	Description
GET	`/map/navigation`	Calculate directions between locations
GET	`/map/distance`	Provide travel distance and time for a route
GET	`/nlp/tts/voice`	List available voices, e.g. language, gender
GET	`/nlp/tts/voice/{id}`	Get information about voice by ID
POST	`/nlp/tts/synthesize`	Synthesize text-to-speech

Call Center
Multiple call centers are used for a Safety & Security implementation. The *emergency call center* is specifically used for emergencies and the automated accident notification service, and has specific interfaces to emergency response units related to the location of the incident. Response times and handling of all types of exceptional scenarios around these services are a top priority. A few difficult examples created by the fact the vehicle was in an accident include getting a phone call without data or getting data without a phone call. In some cases, only parts of data or calls are received and these scenarios will need to be handled as well.

The more standard type of call center handles *information calls* like the roadside assistance service and any other informational inquires a driver may have for assistance with unlocking doors or a stolen vehicle. Additionally, the OEM may support part of the information calls that are related to the vehicle, the brand, and the OEM specifically. The call center is the main *human machine interface* (HMI) for the TSP, so all of the functions mentioned earlier act on at the call center.

The APIs typically start by receiving the call and data from the vehicle, and accessing the customer information. Actions include dispatching emergency services, unlocking the car door, contacting police about a stolen vehicle, and logging the history and interaction with the customer. Example operations regarding emergencies and information calls are:

Action	URL	Description
GET	/account/{id}	Get customer information by ID
POST	/emergency	Connect to local fire department, police or ambulance
POST	/vehicle/{id}/interaction	A request is coming from a customer by VIN
GET	/vehicle/{id}/interaction	Get log of a customer call by ID
GET	/map/place	List of places based on a user's location
GET	/map/place/{id}	Provide detailed information for a specific place
POST	/vehicle/{id}/service	Instruct vehicle to perform action

2.2.5 How Are These Services Used by Customers?

In Generation 1, it started when a customer bought a vehicle, then one of the packages for Safety & Security services needed to be selected from the dealer. Depending on the OEM, there may have been one to three different options. This is managed by the IoT layer and either the customer or the dealer selects these from a portal. Once the customer has selected the package, the TCU is activated by the carrier in the IoT layer and connected to the network. This is very much like the process we see today when we go to a Verizon store, buy a smartphone and get it activated.

With the customer information, the monthly or yearly billing cycle will then be set up and the services can now be used. A good service to test that everything is working is an information call, where we can press the button in the vehicle and talk with a call center to ask about directions or get help with the vehicle. Another fun one to test are the remote functions, remember this is before smartphones, we can call the call center and have them send a signal to unlock the vehicle doors, a great example for a useful vehicle API. Testing automated accident notification is not a good idea, here we need to trust the OEM that it will work in an accident.

Most of the other scenarios are typically event-driven, meaning something has to happen with the vehicle to see a result. Take remote diagnostics, the first time the check engine light comes on, the driver will be alerted and the information about the problem will be shared with the driver. Similarly, with stolen vehicle tracking we usually hope not to use this, but once we have a police report we can call the call center and they will work with law enforcement to track the stolen vehicle and return it safely.

2.2.6 How Far Is It from 'My Cognitive autoMOBILE Life'?

We are assessing the Generation 1 platform by the foundation of the six self-enabling themes (see Sect. 1.6) and illustrate the score in Fig. 2.8:

Self-integrating   1, for getting started with simple GPS data collection and sharing for stolen vehicle tracking.

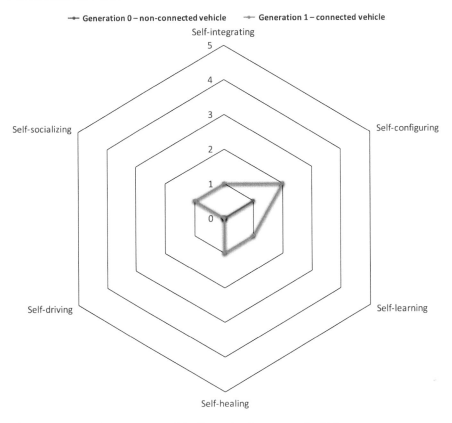

Fig. 2.8 Self-enabling assessment of the Generation 1—connected vehicle

Self-configuring	2, for electronically adjustable settings to customize AUTOmobile components combined with a set of services.
Self-learning	1, for getting started with TTS for voice navigation system with audio navigation instructions.
Self-healing	1, for the simple ability to identify a problem remotely and assist the driver through a call center.
Self-driving	0, for no automation at all.
Self-socializing	1, for getting started with SMS to transmit some data from the vehicle to the Safety & Security services.

2.2.7   How Relevant Is It Across Industries?

The lessons learned for other industries are to take the first step and get connected, the services provided may not be the same, but the ability to be connected to the device or vehicle in this case is the foundation for the cognitive life. Also, the initial focus on the customer or a service for the customer was the foundation to get connected and will lead to more scenarios in the future.

The initial business model was subscription based, we will go into this in more detail in Chap. 3, but basically the customer paid a monthly fee for the services rendered. As we have learned and seen in other industries, sometimes the first step is the most difficult, and aligning that with the correct business model is a challenge as well.

Most companies, including automotive, are really not set up to create connected solutions, mainly because the solution spans organizational boundaries. Typically, product engineering doesn't work with IT on product development, the quality team gains a lot of value, but doesn't determine the requirements to get the data they need. Similar to quality, marketing shares the demand with product engineering for features, but does not own the design and requirements.

These along with measurement systems are a few examples of what makes it difficult to create a connected solution, which is why it typically takes C-level support to launch a connected solution with new values, measurements and outside the normal organizational boundaries. Looking back the mobility revolution [10], understanding the organization at an OEM is very important, and ultimately may require organizational changes to implement the first digitally connected products. It may start with matrix teams to bring together the different skills required to get products connected and bring value to the customers. Some OEMs will keep ownership in engineering, combining skills from IT and their product teams, working directly with the line of business that's selling the product and services in the market. We haven't seen one organizational structure that always works, but getting support from C-level executives that are thinking beyond day-to-day work and individual measurements usually help make it happen, one way or another.

That is well summarized in *Conway's law*, concluding the basic thesis in [3]:

> Organizations which design systems . . . are constrained to produce designs which are copies of the communication structures of these organizations

Other organizational issues are a true understanding of the data and capabilities of a device versus a business or marketing understanding. Due to education and backgrounds, typical technical terms don't mean the same to different people, so a certain functionality can strongly depend on the device supporting that function or capability. A simple example is to understand how often a person in the vehicle has the radio on and to what stations they listen. Theoretically, it seems this would be a simple thing to understand and receive from the vehicle, but technically we cannot get this information about usage electronically without adding additional sensors and costs. Only an example, but marketing thinks one way, engineering thinks another, working to get them aligned is important. To do that, starting with a focus on customer experience is what will drive the most success.

Examples from other industries about taking the first step include connected appliances. A lot of people claim they don't really need a smartphone application to turn on the air conditioner before they get home, but it's not just about the application, it's about the connectivity of the air conditioner, so in the future we can buy filters automatically and do maintenance transparently only when it's required. Another great example is Nest Labs with their first connected thermostat. Who would have thought that a 20-plus-year-old device would be replaced with a $300 connected one. The simple application was remote functionality, learning when people enter and exit rooms, and provide analysis to be more energy efficient. In both these cases, we are starting to see the home automation industry take off and build upon not just having one or two devices connected, but also creating a home automation experience that has both product and ecosystem value. Amazon is a great example now of how home automation is even turning into a cognitive life. More on this when we discuss Generation 4. It will be interesting to see how OEMs in the home connect to so many ecosystems but also differentiate. Connect to the Amazon home, connect to the Google home, but maybe connect to the Panasonic home where differentiated functions exist beyond the others.

2.3   Generation 2: Infotainment and Navigation

The services that were most intensively developed as the next generation at first focused on providing a combination of "*info*rmation" and "enter*tainment*"—known nowadays as *infotainment* for short—such as GPS, news services or weather information. But an infotainment system goes beyond local radio, it is most commonly in combination with the navigation system that is used to find directions based on the location in an AUTOmobile.

In this generation, the first capability added to the vehicle is the head unit. Similar to a smartphone, it supports the applications for both infotainment and navigation. Figure 2.9 as an example of a head unit in a vehicle, most vehicles have a similar layout.

> The *head unit* refers to the hardware in the vehicle that provides infotainment and navigation functions.

The head unit is traditional placed in the center of the instrument panel and replaces or integrates the old vehicle functions like the analog radio, CD player and climate controls from the non-connected vehicle—Generation 0.

Infotainment is the new digital version of music, news, sports, weather, etc. while navigation is the digital automation for directions and places to go. In retrospect, it is astonishing that millions of people used to travel by car without a navigation system and still arrived at the intended destination. Admittedly, sometimes a high social price was paid, in the form of disputes between the driver and the navigator: "I told

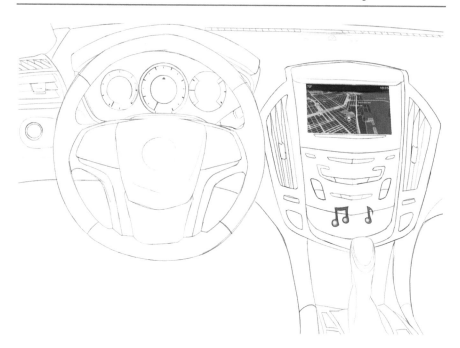

Fig. 2.9 Example head unit gives the user control over the vehicle's infotainment and navigation

you to turn right!" Today, this command comes from the navigation system and thus we have a neutral third party called the head unit, which we can blame.

2.3.1 Solution Architecture

In context of layers, the solution architecture (see Fig. 2.10) stayed the same as Generation 1, but new service domains and capabilities are added to support both the driver and internal uses for the data at the OEM. The IoT entity stayed same as Generation 1, apart from carriers improving the technology to provide the ability to send more data at faster speeds. The big changes occurred in the cloud were mobile, integration, analytics, big data and event management capabilities were developed.

Most of the new capabilities introduced by Generation 2 are in the vehicle and mobility cloud layers. The infotainment applications required both the vehicle and the cloud to integrate multiple sources of information, examples include content from news providers, streaming music, or information from the OEM on recalls or vehicle information. Mobile was introduced as smartphones became popular for simple functions like unlocking the car door or remotely starting the vehicle to warm it up in on a cold morning. Analytics and big data were used internally by OEMs first for collecting and storing all the data from the vehicle and then for analyzing the data to improve quality, reduce warranty claims and help drivers and dealers diagnose

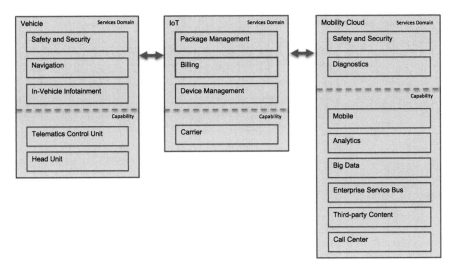

Fig. 2.10 The solution architecture of service domains and capabilities for Generation 2—infotainment and navigation

problems with vehicles. As the data and applications increased, event management was added to transform the data from its binary engineering form to a more human readable form and to orchestrate where and how the data would be used.

2.3.2 Vehicle

In the vehicle layer, we add two new service domains

1. Navigation to provide directions,
2. and In-vehicle Infotainment to provide both information and entertainment.

The core capability is the head unit to give the user control over these two new service domains.

Navigation

An automotive navigation system provides one of the most common sets of services in the connected vehicle, driven by turn-by-turn directions. First provided in the vehicle through an expensive integrated head unit as part of the AUTOmobile controls, then made famous as a cheaper third-party aftermarket add-on by Garmin and later as free app by Google on the smartphone. Initial issues included the maps becoming outdated and the map data having to be loaded manually from DVDs every few years. *Points of interest* (POI) were static, and in later generations both of these could be updated wirelessly, a technology called *over-the-air* (OTA).

A navigation domain provides the following services:

Turn by turn: Directions are the main functions of navigation systems, and provide real-time directions to a selected POI or address entered into the system. The real-time directions are displayed visually and accompanied by voice instructions, and can change based on real-time traffic or weather.

POI: A POI is a specific place that someone may find useful or interesting, most commonly food, lodging, gas stations, parking, hospitals, parks. etc. POIs are shown and used in the navigation system and provide a possible destination for travel, or even just a new interim destination at any time during the trip. Most navigation applications allow users to set their own POI for common places they travel to like work, school or home.

Traffic & weather: The most common dynamic information shared with in-vehicle applications is traffic and weather. They are displayed as graphics on the head unit or provided as audio so the driver can adjust the route according to any specific incidents. Vehicles also provide GPS data to traffic providers who aggregate vehicle data and other sources from cities to estimate current traffic conditions and their impact on the arrival time.

Geofencing: Geofencing is a service that provides a digital fence around a defined area. A user can set a geofence for a specific area to get notified when a vehicle leaves or enters the area, useful parental controls to track their children during those teen years. Geofences are also used for defining areas for weather, POIs, and traffic-related information.

Example operations regarding navigation systems are:

Action	URL	Description
GET	/map/navigation	Calculate directions between locations
GET	/map/distance	Provide travel distance and time for a route
GET	/map/place	List places based on a user's location
GET	/map/place/{id}	Provide detailed information for a specific place
GET	/map/suggest	List suggestions for address search
GET	/vehicle/{id}/fence	Get polyline geo-fence
POST	/vehicle/{id}/fence	Create polyline geo-fence
DELETE	/vehicle/{id}/fence	Delete polyline geo-fence

More comprehensive APIs are well known by Google Maps and Places APIs.

In-Vehicle Infotainment

The service domain

> *in-vehicle infotainment* (IVI) is a set of mobile or in-vehicle services that provide both information and entertainment, typically related to traveling and improving the driver experience.

This domain provides the following services:

Music:	A user selects an input source for music to be played in the vehicle. Radio and music have a long history in automotive, but what infotainment brought to the vehicles was similar to smartphones, and digital music. Both steaming and satellite radio are popular, SiriusXM, Pandora, iHeart Radio, Aha, are all common providers in the US. Digital also included podcasts, books etc., all related to information and entertainment while driving.
Digital control:	A user selects a digital icon to operate a vehicle function. As IVI and navigation created the demand for a digital display in the vehicle, OEMs started to move more functions from manual control to digital control, including climate, fan speed, temperature, audio, and vehicle preferences.
Bluetooth integration:	A user pairs a Bluetooth phone with the vehicle and can operate hands-free phone functions during driving. Two functions are provided in the vehicle based on Bluetooth integration, hands-free and streaming audio. The hands-free function supports both voice calls and texting using only the in-vehicle speakers and microphones, streaming audio allows users to play content from their smartphone through the vehicle's speakers.
In-vehicle apps:	A user selects an icon to launch a specific application. Similar to the smartphone, the head unit has the capability to support applications, OEMs allowed outside suppliers to develop specific applications for the vehicle. Functionality included streaming audio, sports, news, weather, traffic, parking, fuel prices, etc. The main difference between in-vehicle application and mobile integration is that the application runs on the head unit and not on another device. An entire chapter could be dedicated to the current debate on in-vehicle versus mobile applications, but for this book, we will not focus on the debate, but on the fact that both will exist moving forward.

This service domain is a very rich area of continuously newly developed APIs. Some example operations regarding IVI are:

Action	URL	Description
GET	/music/artist	List several artists
GET	/music/artist/{id}	Get artist details by ID
GET	/music/categories	List of music categories
POST	/account/{id}/music/playlist	Create a user's music playlist
GET	/account/{id}/music/playlist	Get a user's music playlist
PATCH	/account/{id}/music/playlist	Update a user's music playlist
DELETE	/account/{id}/music/playlist	Delete a user's music playlist
GET	/account/{id}/music/recommend	Recommendations based on seeds
GET	/account/{id}/device	List of a user's devices, e.g. Bluetooth
GET	/account/{id}/device/{id}/service	List of services discovered for device
GET	/account/{id}/map/trip	List of a user's trips
GET	/account/{id}/vehicle	List of a user's vehicles
GET	/vehicle/{id}	Get vehicle data by VIN
GET	/vehicle/{id}/statistic	Get statistics, e.g. recent trip
GET	/vehicle/{id}/status	Get vehicle status, e.g. doors locked

Head Unit

A head unit is much like a smartphone. It has a display, a CPU, a modem, and applications. In addition, it has like the TCU with a connection to the CAN bus system. The CAN bus, much like the Ethernet network in an office building, gives the unit access to vehicle information, vehicle control and vehicle sensors, mostly for sound and volume controls.

The CPU runs the infotainment and navigation applications along with the other in-vehicle applications. A core capability included to support most applications—not only navigation as in Generation 1—was voice-recognition software and the ability to just say voice commands instead of pressing mechanical buttons or turning knobs. The voice commands were pretty basic and worked in a command tree structure, meaning we could only say certain commands at certain times. An important function to support hands-free operation and limit driver distraction, ultimately improving safety, but an area that has caused OEMs a lot of headaches and bad press regarding usability. More on that later.

An optional modem provides connectivity for all applications and the in-vehicle Wi-Fi. With the head unit and its smartphone-like capabilities, embedded APIs and in-vehicle development expanded greatly. This is not intended to be an extensive list of APIs, but just enough to provide an understanding of how these new functions were implemented.

Navigation APIs focus on the destination and display the map with turn-by-turn directions. Advanced head units would display weather and traffic at the same time, overlaying the view of where the driver was going. Geofencing has many uses, as we discussed, but is simply based on a location, a defined area and the type of alert the user wants to set.

Diagnostics APIs improved, more data was sent along with the DTC including a buffer of selected parameters before and after the DTC occurred. As in Generation 1, the check engine light is the most common notification a driver has for an issue with the vehicle. But instead of just a malfunction indicator lamp, the head unit has much more capabilities to provide understandable information to alert the driver when something has gone wrong with the vehicle operation.

Mobile integration for remote functions introduced new APIs, but used the APIs already in place by the call center. Smartphone integration for Apple CarPlay and Android Auto required a connection to the head unit and the remote display feature. The remaining functions were handled by the application on the smartphone. Bluetooth also required connection APIs to support the functions for making phone calls or streaming audio to the loudspeakers. The digital control APIs were replacements or duplicates of existing physical functions and music controls involved selecting from the main different channels and playing the selected song.

The capability of a head unit, similar to IVI, is a very rich area of continuously newly developed APIs. Some example operations regarding the head unit are:

Action	URL	Description
GET	`/map/navigation`	Calculate directions between locations
GET	`/device/{id}/map`	Display map on head unit
GET	`/device/{id}/map/weather`	Display weather info on head unit
GET	`/device/{id}/map/traffic`	Display traffic info on head unit
GET	`/device/{id}/map/fence`	Display geo-fence on head unit map
POST	`/device/{id}/map/fence`	Enter geo-fence on head unit map
DELETE	`/device/{id}/map/fence`	Delete geo-fence on head unit map
POST	`/device/{id}/map/navigation`	Enter travel destination
POST	`/device/{id}/map/distance`	Display travel distance, time for route
POST	`/device/{id}/map/place`	Display places based on user's location
POST	`/device/{id}/map/place/{id}`	Display details for a specific place
GET	`/device/{id}/diagnostic`	Display readable diagnostics on head unit
POST	`/account/{id}/device/{id}/connect`	Connect device to head unit

(continued)

Action	URL	Description
DELETE	`/account/{id}/device/{id}/connect`	Disconnect device from head unit
POST	`/account/{id}/device/{id}/resource`	Device uses head unit resource, e.g. display
DELETE	`/account/{id}/device/{id}/resource`	Device terminated use of head unit resource
POST	`/account/{id}/device/{id}/service/{id}`	Instruct device to perform action, e.g. phone call
GET	`/nlp/tts/voice`	List available voices, e.g. language, gender
GET	`/nlp/tts/voice/{id}`	Get information about voice by ID
POST	`/nlp/tts/synthesize`	Synthesize text-to-speech
GET	`/nlp/stt/model`	List available models, e.g. en-US broadband
GET	`/nlp/stt/model/{id}`	Get information about model, e.g. min. sample rate
POST	`/nlp/stt/recognize`	Send audio to get transcription results

2.3.3   Mobility Cloud

In the mobility cloud layer, we added the new services domain 'Diagnostics' to optimize maintenance. In addition, we defined five new capabilities:

1. Mobile applications connecting the smartphone with the vehicle,
2. Analytics to understand what and why something happened and to better support driver safety,
3. Big data to collect and combine information from different sources,
4. Enterprise service bus to integrate heterogeneous systems, and
5. Third-party content for value-added services related to the connected vehicle.

Diagnostics

First introduced in Generation 1, diagnostics capabilities continually advance with every generation of connected vehicle. OEMs introduced health reports, emails that summarized the health of the vehicle and remind us of upcoming maintenance. More advanced functions included predictive diagnostics were data collected from the vehicle is used to better predict when maintenance is needed. Better remote diagnostics capabilities are available, so when a critical situation occurs, the remote operator can better diagnose the problem and get it resolved as quickly as possible or schedule an appointment at the preferred dealer.

Example operations regarding diagnostics are:

Action	URL	Description
GET	/vehicle/{id}/diagnostic	Get diagnostics report of vehicle
POST	/vehicle/{id}/diagnostic/log	Log DTC for vehicle
POST	/vehicle/{id}/diagnostic/predict	Predict vehicle maintenance

Mobile

The first type of mobile applications for OEMs was to extended Generation 1 capabilities to the smartphone. Most common is door unlocking, other features included honking the horn to locate the car in a parking lot or flashing the lights for the same reason.

Displaying diagnostics functions on smartphones along with basic statuses of the vehicle became common as well, especially with electric vehicles. The status in electric vehicles included the remaining battery charge, the estimated range, and scheduling for charging the vehicle at home.

Location functions along with planning routes and then sending them to the vehicle were developed as well. As smartphones gained popularity and as more and more applications became available, OEMs were under pressure to integrate those capabilities into the vehicle. Two major approaches exist in the market today, SmartDeviceLink[4] and Apple CarPlay/Android Auto. SmartDeviceLink allows any application to integrate specific functions from their application with vehicle controls on the steering wheel, voice enablement, and graphics on the head unit. It is limited in what can be displayed from an application, but the main advantage is the same application that was downloaded from an app store to a smartphone can be used in the vehicle if programmed with SmartDeviceLink.

Apple CarPlay and Android Auto basically work based on the concept of a remote display, meaning what we would see on the smartphone is now displayed on the head unit screen. One disadvantage is the application HMI does need to change to meet OEM driver distraction standards. The advantage is we can have a richer HMI experience with the application similar to how we use it on our smartphone. An entire chapter could be dedicated to the current debate about smartphones taking over the vehicle applications, but for this book, we will not focus on that debate, but on the fact that these two options will exist moving forward.

---

[4]"SmartDeviceLink" http://smartdevicelink.com. Accessed: July 26, 2017.

Example operations regarding mobile are:

Action	URL	Description
GET	/vehicle/{id}	Get vehicle data by VIN
GET	/vehicle/{id}/statistic	Get statistics, e.g. recent trip
GET	/vehicle/{id}/status	Get vehicle status, e.g. door locked
GET	/device/{id}/diagnostic	Display readable diagnostics on smartphone
POST	/vehicle/{id}/service	Instruct vehicle to perform action
POST	/device/{id}/send/device/{id}	Send data, e.g. smartphone map to head unit

Analytics

In Sect. 1.2.1 we introduced analytics, but in the previous Generation 1, we did not list it as a core capability, because it was not used in the scope of the kind of connected vehicle summarized there. However, in Generation 2 we got on the descriptive level (see Fig. 1.8) mostly driven by vehicle diagnostics to understand what and why something happened and to better support driver safety, e.g. through the call center.

Beyond vehicle diagnostics, spatial analysis is one of the most common features now in automotive that is used to analyze location and map related information. We have talked about traffic data and learning driving patterns before, both of them fall into the spatial analysis category.

Example operations regrading analytics are:

Action	URL	Description
GET	/analytics/query	List named queries
POST	/analytics/query	Create a named query
GET	/analytics/query/{id}	Get information about a query
DELETE	/analytics/query/{id}	Delete a named query by ID
POST	/analytics/query/{id}/execute	Execute a named query
GET	/analytics/query/{id}/result	Get executed query result
DELETE	/analytics/query/{id}/execute	Stop a query execution
GET	/analytics/query/execute	List query execution
GET	/account/{id}/music/recommend	Recommendations based on seeds

Big Data

Big data refers to data sets that are so large or complex that traditional data processing methods are inadequate to handle them.

However, note the difference between typical structured types of data stores used in the industry which are relational databases, followed by time-series databases, and finally the introduction of Hadoop[5] for semi-structured and unstructured data as what we know as the fundamental tools for big data today.

We won't focus on the nonfunctional requirements such as real-time aspects, security and privacy. Then we can reduce the core functions of big data management to ingest, enrich, store, and explore (see Fig. 2.11):

1. Ingest raw data refers to data collection, also popular in the industry as the IoT, which has many definitions,. We will use the more narrow scope of the ability to connect and collect data from devices, as data types, structured, unstructured and semi-structured with all the examples like text, video, images etc.
2. The enrichment of the raw data is important to improve the quality and relevance for the targeted scope to be extracted from the data. Privacy may become an issue when multiple sources will be integrated into a bigger personal context of the driver.
3. The data stores like operational and data warehouse are important for certain types of data and whether they are used for an application or for off-line analysis. Popular in automotive are traditional data sources defined for geographical data, sensor data and smart devices.
4. Finally, the value-added content will be created by exploring the data through the analytics capabilities we described previously.

Early lesson learned: Do not collect data for the sake of collecting data and then hope to simply find value in it. Initially, when it was expensive to send data, engineers spent a lot of time designing and understanding the data they wanted.

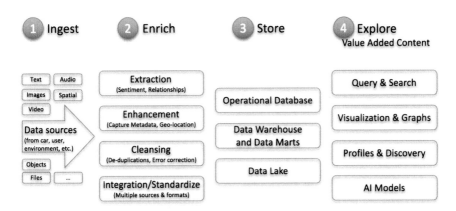

Fig. 2.11   The core functions of big data management include ingest, enrich, store, and explore

[5]"Apache Hadoop" http://hadoop.apache.org. Accessed: July 26, 2017.

Now with technologies like WiFi, some engineers no longer think and send more data then they need, or in a simple case data that is not even changing over time. A difficult trade-off that sometimes just needs some basic analysis before gigabytes of data a day are sent to a cloud. For example, a vehicle today can generate almost "2 petabytes of data per car per year".[6] That doesn't mean we need to pay the carrier to collect it all. Note, we have seen engineering examples where this amount of data is collected, but not in the normal operations of a vehicle so far.

One other factor is speed. The improvement in wireless technologies in the IoT, along with the cost, made certain scenarios more viable. Data storage, similar to wireless, speed and cost of storing data are the driving factors and in the cloud, all types of options are available and certainly related to the term big data, the amount of data we can store and analyze has certainly improved over the years. In line with data storage, curation or the plan to understand how to manage data over time becomes more important with the amount of collected data. Data models are the abstract version of the collected data, which logically helps to set up how the data is organized and how people will use it.

Last but not least, data security and privacy, but we will talk about this more in Chap. 4 as we look at new technologies and approaches for consumers to protect the data about their person or their vehicles. We don't want to go into all the methods and approaches for data security and privacy, but it's a fundamental requirement that concerns OEMs and customers the most.

The best internal example of big data collection and analytics is warranty cost reduction. A solution that takes data from multiple sources, first the warranty claims for the dealer, second real-time diagnostic data from the vehicles, and third the service records from the dealers. The biggest value comes from the real-time data from the vehicle. If we can correlate this data with problems at the dealerships, it can give us an earlier warning of troubles ahead. When a warranty problem leads to a recall, the time it takes to identify the problem is the most costly, so shortening that time is very valuable. Predictive algorithms on real-time data can also be used to help consumers or fleet managers see a problem before it happens and get the routine maintenance done to avoid downtime in the field.

The best external examples of data used outside the OEMs are PAYD insurance and traffic data, both are based on collecting and selling GPS data from vehicle. Everyone has seen the traffic patterns either reported on Google or the morning news, where a green line represents a good traffic flow, yellow is slow and red is stopped. The foundation for that analysis is the big data collected from vehicles. PAYD insurance uses similar data, except the analysis is on the driver, typically giving the driver a score that translates into the insurance rates that need to be paid.

Data APIs can be relatively straightforward for connected vehicles. At this point, most applications and content are outside the OEM, so the data from the vehicle

---

[6]"Self-driving Cars Will Create 2 Petabytes Of Data, What Are The Big Data Opportunities For The Car Industry?" https://datafloq.com/read/self-driving-cars-create-2-petabytes-data-annually/ 172. Accessed: July 26, 2017.

and about the customer are the key data elements. The TCU or the head unit will send location and event type information, along with vehicle sensor information and diagnostics information.

Example operations regarding big data are:

Action	URL	Description
GET	`/vehicle/{id}`	Get vehicle data by VIN
PATCH	`/vehicle/{id}`	Send vehicle data by VIN
GET	`/analytics/dataset`	List all datasets
POST	`/analytics/dataset`	Create a new dataset
GET	`/analytics/dataset/{id}`	Retrieve information of a dataset
PATCH	`/analytics/dataset/{id}`	Update information in a dataset
DELETE	`/analytics/dataset/{id}`	Delete a dataset

Enterprise Service Bus

> An *Enterprise Service Bus* (ESB) implements an interaction and communication system between distributed software applications in an SOA. It handles different types of messages and protocols within the system and routes requests to the appropriate destination.

In Sect. 1.3 we established that an ESB is not necessary for the microservices style of architecture to focus on the endpoints and the associated customer experience. But the ESB capabilities can get very essential for integrating the multiple channels supported across the vehicle, mobile and web. As the amount and complexity of the data from vehicles increased and additional applications related to the data were added, the integration with all the different services and systems was required.

First was the need for more data conversions so more application could use the data, along with data orchestration to support the more complex rules, queueing, and flows of the information between internal and external systems. A simple example of an ESB implementation is the collection of GPS data from the TCU. The data are first processed by the ESB and stored in a queue to ensure they are not lost, then converted from a binary format to a human-readable format for latitude, longitude, speed and direction. Since multiple systems are typically interested in the location information, the ESB handles sending it to all the different systems.

As we have been discussing APIs, an ESB is used as the facilitator between APIs so that the rules and business logic are separated from the applications. API management was also implemented to help control and support many different

Fig. 2.12 ESB pattern is integration-centric and API management is consumption-centric

services and versions of these services over time from a consumption-centric perspective. API management also provided a level of security so external providers could more easily access data within an enterprise. Hence, the ESB is a back-end integration system and should not be confused with an API gateway that is focused on making the APIs secure for consumption (see Fig. 2.12).

We will define two types of integration: one is internal integration where information is used internally to the OEM, and external integration were information is shared with third-party ecosystems. External integration is very common for both the head unit and call center capabilities we defined earlier. Most of the in-vehicle applications access information from third parties, like news, traffic, weather, etc. Most of the integration points are fixed while some are dynamic, meaning different suppliers can provide similar data. The call centers are closely integrated with map providers, POI providers and for Safety & Security with a public safety answering point provider which associates locations with the responsible emergency responders.

If a TSP is used by the OEM, then integration points will exist to obtain information about the vehicle, when it was built, what components are in it and when it was sold. OEMs' engineering, manufacturing, and CRM systems typically contain this information. Internal integration is mainly with quality, service, engineering and marketing organizations. Marketing is interested in the customer and the features customers are using in the vehicles. Quality and warranty organizations like to collect vehicle and diagnostics data, and for the service organizations focusing on scheduling dealer appointments and improving the overall service experience are becoming important.

Some example operations regarding ESB are:

Action	URL	Description
GET	/namespace	List ESB namespaces
POST	/namespace	Create an ESB namespace
GET	/namespace/{id}	Get ESB namespace by ID
PATCH	/namespace/{id}	Update an ESB namespace
DELETE	/namespace/{id}	Delete ESB namespace by ID
POST	/namespace/{id}/rule	Create a namespace authorization rule
GET	/namespace/{id}/rule/{id}	Get a namespace authorization rule
PATCH	/namespace/{id}/rule/{id}	Update a namespace authorization rule
DELETE	/namespace/{id}/rule/{id}	Delete a namespace authorization rule
POST	/namespace/{id}/topic	Creates a topic in namespace ID
POST	/topic/{id}/rule	Creates an authorization rule for topic ID
POST	/topic/{id}/subscription	Create at topic subscription
GET	/topic/{id}/subscription	List subscriptions under topic ID
PATCH	/topic/{id}/subscription/{id}	Update a topic subscription
DELETE	/topic/{id}/subscription/{id}	Deletes a subscription from topic ID

More comprehensive examples exist from Microsoft Azure Service Bus REST APIs.[7]

Third-Party Content

Content from third parties includes map data, traffic data, POI data, news, weather, sports, music, etc. In Generation 2, most of the data becomes more dynamic, with applications being added and additional connectivity more data can be presented in a dynamic fashion. Companies like Inrix, SiriusXM, and TomTom are the leaders in bringing dynamic content into vehicles.

In Chap. 4, we will discuss more how the business model, personalization, and value of this information is changing the industry. Probably the most common area for APIs and implementation of ESBs would be related to the integration of content providers. Either from the head unit directly or from the enterprise, common APIs for news, sports, weather, traffic, POIs, fuel, etc. would be used and often replaced as new partnerships emerge or suppliers change.

---

[7]"Azure Service Bus REST API" http://docs.microsoft.com/en-us/rest/api/servicebus/. Accessed: July 26, 2017.

Example operations regarding third-party content are:

Action	URL	Description
GET	/news/source	List the available news sources and blogs
GET	/news/source/sport	List the available sport news sources and blogs
GET	/news/article	List live article metadata from a source
GET	/traffic/incident	Get traffic incident information
GET	/traffic/flow	Get traffic flow information
GET	/weather	Get current weather by location
GET	/weather/forecast	Get weather forecast by location
GET	/weather/history	Get historical weather data
GET	/weather/pollution	Get air pollution by location
GET	/map/place	List places based on a user's location
GET	/map/place/{id}	Provide detailed information for a specific place
GET	/map/place/category	List available place categories for a location

2.3.4   How Are These Services Used by Customers?

With the introduction of the head unit into the vehicle, OEMs gained an important platform to control information and interactions with the customer. IVI went beyond the traditional radio and compact disc players, and allowed users to bring digital content into the vehicle and started to customize the user experience. Originally, some head units even stored content, just like traditional music players. But nowadays the connectivity between a user's device and vehicle is more seamless with wireless streaming or simple audio plugs. Bluetooth and hands-free became very popular, with some states even putting laws in place that require it. Bluetooth even allows a person walking and talking on a smartphone to enter a vehicle and the sound and microphone are automatically transferred to the vehicle so a user can continue his conversation seamlessly. A great example where the Bluetooth implementation focused on the customer experience. Other Bluetooth features included, hands-free dialing and answering while driving.

Basic applications like navigation became popular, followed by other forms of entertainment beyond music, including Internet radio, satellite radio, and other media for news, weather and sports. As navigation grew in popularity, it moved from just destinations to integrated traffic and weather or to specific POIs for parking and hotels. Navigation systems weren't always integrated into the vehicle, Garmin was one of the first to introduce a low-cost aftermarket device that consumers attached to their dashboard. Google was the first to introduce navigation on a smartphone that also allowed consumers to bring the experience to their dashboard. Later with Android Auto, the navigation application was one of the first to be projected or remotely displayed on the head unit. As connectivity improved, more and more of the navigation systems went to dynamically updating the information like POIs and in some cases the maps themselves. This is an important feature to watch in the future, similar to the smartphone. The more personalized and dynamic we can design things, the better the customer experience.

Back to the Garmin. We remember plugging into our computers and downloading updated maps and loading them onto the unit. Vehicles worked in a similar way. Typically the dealer would load discs with the updated data into the car. As more and more applications and integration occurred, the customer experience became more and more important, even to the point where it started to affect vehicles sales. A big turn in the industry occurred when Ford announced[8] increased vehicle sales due to Ford Sync. Technology and customer experience were now really affecting the vehicle. OEMs have struggled with applications in the vehicle and now with Apple CarPlay and Android Auto the debate will continue, but clearly creating the best customer experience will win out.

2.3.5 How Far Is It from 'My Cognitive autoMOBILE Life'?

We are assessing the Generation 2 platform by the foundation of the six self-enabling themes (see Sect. 1.6) and illustrate the score in Fig. 2.13:

Self-integrating   2, for the introduction of digital content to the vehicle and smartphone for remote functions drove the improvements.

Self-configuring   3, for the IVI starting to replace mechanical functions with digital functions along with support for more settings to be personalized.

Self-learning   2, for the ability to plug intelligence into the vehicle via smartphones, but still mostly descriptive analytics.

Self-healing   2, for improvements in remote diagnostics that helped the customer directly and the OEMs internally.

Self-driving   1, for getting started with driving assistance functions visualized inside the IVI.

Self-socializing   2, as the vehicle started sharing its location and speed to support traffic data and the in-vehicle as well as the smartphone applications provided the driver even more social capabilities.

2.3.6 How Relevant Is It Across Industries?

The lessons learned for other industries are understanding the HMI, data to support internal business processes, and understanding the platform around the customer device and the value it brings.

---

[8]"Ford Sees Sales Increase Thanks to Its SYNC and MyFord Touch" http://www.inautonews.com/ford-sees-sales-increase-thanks-to-its-sync-and-myford-touch. Accessed: July 26, 2017.

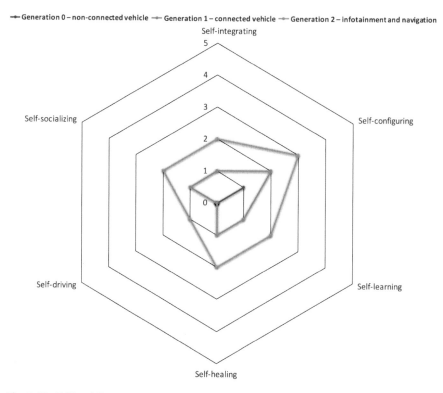

Fig. 2.13 Self-enabling assessment of the Generation 2—infotainment and navigation

Let us begin with the HMI. The industry started off struggling in this area, improved a bit, but unfortunately could not keep up with the usability standards created by smartphones. Looking at the automotive and mobility industries, there are no standards for analysis and the results of HMI design are mostly subjective. However, compared to smartphones, the automotive industry has come a long way, but still has a ways to go. First no touchscreens, then finally the touch screen is available, next no swipe, then finally swipe is available, next limited commands for voice recognition which are hard to remember, and then some improvements with NLP for voice recognition.

Looking at the physical characteristics of IVI, which is the mix between manual and touchscreens, most OEMs are still struggling with this, some committing to all digital while others are trying to find the correct mix. Also the mix of functions of the IVI versus on the instrument panel has also caused confusion on how to execute certain functions. This is caused by multiple vendors supporting the different areas of the vehicle design and really not working together, which doesn't create a standard user experience. The initial integrations of smartphones into the vehicle were clumsy, Bluetooth standards certainly helped, but now both Apple CarPlay and Android Auto show a much better HMI as well as a better overall customer experience. We just talked extensively about HMI in vehicles, but what it shows is the importance of getting it right and understanding when to release functions or not. We are sure that Apple could have had Siri out a few years earlier, but waited until they felt it and users were ready makes for a better overall experience. The HMI and functions are even said to affect which vehicles customer will purchase, so as companies enter the IoT, think HMI and don't underestimate its importance. We will discuss this more in the next section on the next-generation cognitive vehicle and the growth and change we will see in areas like HMI and customer experience.

Next, based on the increase of bandwidth and cost for data, collecting data from the vehicle and using the data for internal and external value is one of the key value drivers for all IoT solutions. Learning to design for collecting this data and changing the business around this data is as important as the product itself. One of the struggles the automotive industry had was that the engineers design for the functionality in the vehicle, not for the functionality of the data in their organization. And like most designs, if we don't take it into consideration, it won't happen. There are OEM-built solutions that don't even give them a realistic view of the customer and the functions they are using on a daily basis, strictly because it wasn't included in the design. Remember the importance of the data for warranty, marketing, and quality, but collecting data and doing nothing with it can be just as bad. Having the people in the organization to deal with the new data and the new processes is as important as the product itself.

This leads to the third and final point regarding the platforms supporting the device, both on the device and externally. OEMs again have struggled to understand how to support an application platform in the vehicle. Only recently have OEMs made announcements regarding open software development kits for vehicles in comparison to SmartDeviceLink, which only supports applications on the smartphone integrated into the vehicle. Certainly, cost was an issue. Application developers did not want to support so many different types of head units and operating systems, but the initial lack of a strategy of how to do this led mostly to confusion, limited capabilities and a bad customer experience overall. Remember that the device platform and ability to use data in an ecosystem, both internally and externally, are important design points for IoT solutions. Even the new televisions have better standards for adding applications like Netflix, YouTube, Skype, etc. and imagine Amazon Echo without an ecosystem to add additional products or services to an order. Internally, similar standards and platforms are needed for organizations to use the data. Consumption-centric APIs in combination with big data and an integration-centric SOA are key decisions and costs that should also be considered in initial IoT solutions.

2.4   Generation 2: Commercial Telematics

A short break from the consumer business to touch on commercial telematics in that timeframe of the second generation. We won't do any more generations for commercial, but wanted to introduce the industry because of its large success.

It doesn't have the volume of the consumer business, but it can really show the value and with countless ROI tools available on the Internet, it is not hard to understand why. First launched by Qualcomm in 1988, OmniTRACS was the first vehicle tracking and messaging system for long-haul trucking based on satellite communication.[9] Remember, this is years before cellular was available and Qualcomm was just starting out in the communication industry. We could write a similar book just about this industry, but instead chose to highlight some of the similarities in solution architectures and strategy that drove OEMs and some of their unique mobility solutions. As far as lessons learned go, this industry is certainly the best example of design for customer experience and understanding value that we see in the market today.

Commercial telematics is broken up into multiple industries, each using similar solutions and gaining similar value. Its roots are in long-haul trucking, but also include small and medium-size segments, agriculture, construction, government fleets, service/delivery and utilities. One main difference from the automotive industry was their success in the aftermarket. Not that the OEMs in certain industries did not provide their own solutions, but the majority of the solutions were provided and installed in the aftermarket. In Fig. 2.14 we can see a sample of the aftermarket type solutions used in the industry including in cabs, smartphones, tablets, black boxes and dongles.

Depending on the type of truck and the type of solution, no one solution works for everyone. Aftermarket companies like AirIQ, Fleetmatics, Mix Telematics, TomTom, and Trimble are leaders in the global market, along with hundreds of smaller companies supporting small fleets ranging from one to five trucks.

2.4.1 Solution Architecture

In context of layers, the solution architecture (see Fig. 2.15) required to build a commercial telematics solution has a lot of similarities with the connected vehicle consumer business, but the largest difference is in the type of services implemented in this commercial industry.

2.4.2 Vehicle

In the vehicle layer, we define multiple new service domains

1. Safety and compliance with regulations and protection of drivers and assets,

---

[9]"Qualcomm Celebrates 15th Anniversary of OmniTRACS Satellite Mobile Communications System" https://www.qualcomm.com/news/releases/2003/08/25/qualcomm-celebrates-15th-anniversary-omnitracs-satellite-mobile Accessed: July 26, 2017.

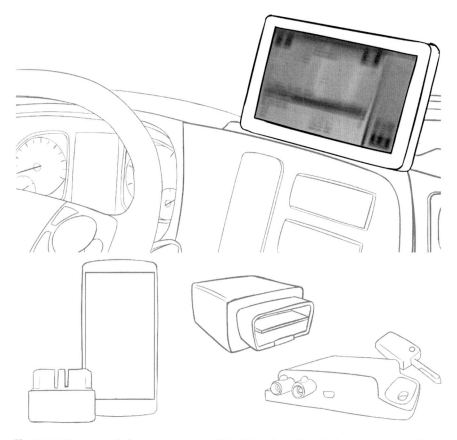

Fig. 2.14 Examples of aftermarket commercial vehicle telematics solutions used in the industry including in cabs, smartphones, tablets, black boxes, and dongles

2. Navigation for optimizing fleet operations and avoiding service/delivery delays and problems, and
3. Schedule and dispatch to plan the delivery or service business,

and two capabilities

1. Telematics control unit to connect vehicle data with the cloud systems,
2. and Head unit to display and control the information and navigation.

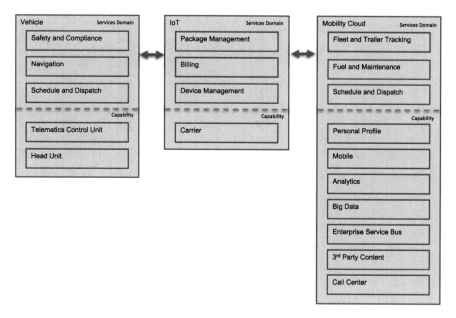

Fig. 2.15 The solution architecture of service domains and capabilities for Generation 2—commercial telematics

Safety and Compliance

The domain related to safety and compliance comprises following services:

DOT compliance: The US Department of Transportation (DOT)[10] regulates the commercial trucking industry and commercial telematics systems electronically assist drivers in keeping track of and comply with these regulations. Hours of service is the most common feature. It requires a driver to log on-duty time, off-duty time, non-driving time, driving time and sleeping time. Europe has mandated electronic recording of hours of service, the US has recently passed a similar law. Another regulation called pre-trip and post-trip inspection requires a driver to inspect the condition of the tractor and trailer for each trip, including information about tires, lights, air hoses, brakes, fluid levels, etc. Fuel tax is actually a state law that requires trucks that cross state lines to report their time and fuel usage in that particular state. The new electronics system automatically provides the driver and fleet operator with most of the above

---

[10]"US Department of Transportation" http://www.transportation.gov. Accessed: July 26, 2017.

	information, saving the driver lots of time compared to the original paper forms.
Event recording:	In case of an accident a TCU can collect information related to the condition of the truck, trailer and environment before and after an accident. Insurance companies and law enforcement can use this information to help determine or validate the cause of an accident. Video devices can also be installed to assist in visually recording events around the time of an accident. This information can save a fleet operator or driver money related to insurance or fines depending on the situation.
Theft prevention:	Theft prevention can include the truck, the trailer or contents of the trailer. All can be monitored and controlled by adding additional sensors to a standard TCU. The sensors can notify the driver or fleet operator when a trailer is disconnected from the truck or when the doors of the trailer are opened. In some cases, even when the truck is stolen, the location can be used by law enforcement to track down the thief. These types of additional functions can help the driver and fleet operator to reduce insurance claims.
Driver scoring:	How a driver performs on the road can save a fleet operator a lot of money beyond safety. A driver's acceleration, deceleration, turning, and braking patterns can be monitored and used to access their driving behavior and reviewed with the fleet operator on a periodic basis. Certain types of driving behavior can save both on wear and tear on the truck and the amount of fuel that is used. Speeding alerts, where the driver's speed is compared to the speed limit stored in the navigation system also save on insurance and fines.

The APIs are going deeper than the driving behavior, the vehicle status and diagnostics that we introduced in the previous sections. Driving APIs are controlling country regulations and logging violations. The vehicle condition is tracked in real time, especially in accident situations.

Example operations regarding safety and compliance are:

Action	URL	Description
GET	/account/{id}/score	Get driving scores based on past data
GET	/account/{id}/vehicle/{id}/log	Get driver log
GET	/account/{id}/violations	Get driving or shift violations
GET	/vehicle/{id}/status	Get vehicle status, e.g. door locked
GET	/vehicle/{id}/condition	Get vehicle condition, e.g. damages

Navigation

The domain related to navigation comprises the following services:

Turn-by-turn: Turn-by-turn directions are very similar to the navigation
                         services for the consumer-oriented connected vehicle. Enter
                         the destination and get route guidance both visually and
                         audibly. Fleets can have additional route requirements based
                         on weights for bridges and heights for overpasses that may
                         need to be avoided. Savings from more accurate routes and
                         not getting lost help with fuel costs and customer satisfaction
                         for on-time arrivals.

Traffic and weather: Traffic and weather are key events that can affect fleet
                         operations and navigation. Improved accuracy and frequency
                         of updates can help fleets to avoid situations that can cause
                         delays and problems in their operations.

The available APIs related to navigation are very similar to the navigation APIs
introduced previously in the connected vehicle platform, but commercial telematics
is more tightly integrated with third-party content like weather and traffic to
optimize the business.

Example operations regarding navigation system are:

Action	URL	Description
GET	`/map/navigation`	Calculate directions between locations
GET	`/map/distance`	Provide travel distance and time for a route
GET	`/map/arrival`	Get estimated time of arrival
GET	`/traffic/incident`	Get traffic incident information
GET	`/traffic/flow`	Get traffic flow information
GET	`/weather`	Get current weather by location

Schedule and Dispatch

The domain related to schedule and dispatch comprises the following services:

Job status: For the fleets that are in the multiple delivery or service
                         business that require the driver to make multiple sched-
                         uled stops in a single day, job status is an important factor
                         to driving profits in their business. Knowing when the
                         driver arrives at a location, when the driver leaves, and
                         that a delivery or service is complete helps the fleet man-
                         ager manage productivity and workload for the fleet. With
                         improved productivity, a service company can get to more
                         customers in a day, increasing the opportunity for more

revenue and with accurate tracking, a fleet can improve customer service and satisfaction by having more on-time arrivals. Messaging capabilities also exist in the system to inform the fleet operator either electronically or by voice about any situations that may affect daily operations.

Schedule management: Beyond knowing the status of the daily jobs, schedule management is another important aspect for delivery and service fleets. Many factors can change a schedule throughout the day; delays on a particular job, parts availability, a customer's availability, etc. can all cause changes in a schedule. What is important is to be able to communicate those changes to the fleet and adjust accordingly. The value again relates to customer service and satisfaction by keeping everyone informed about and changes.

The APIs are related to optimizing scheduling and routing to maximize the number of jobs or deliveries that are known ahead of time. Deliveries can be completed in the day whilst minimizing mileage and balancing the workload across all of the available vehicles. In addition, ad-hoc jobs need to be assigned to the best driver and vehicle to carry them out based on criteria including location, availability, skillset and permitted response time.

Example operations regarding schedule and dispatch are:

Action	URL	Description
GET	/fleet/{id}/vehicle	List available vehicles
GET	/account/{id}/shipment	List shipments planned by driver
GET	/vehicle/{id}/shipment	List shipments planned by vehicle
GET	/vehicle/{id}/route	Get shipments route for vehicle
GET	/account/{id}/stop	Get scheduled driver stops
GET	/schedule/{id}/job	List scheduled jobs and status
GET	/schedule/adhoc	List jobs that need to be assigned

Telematics Control Unit

In commercial telematics, the TCU is similar to the TCU in the connected vehicle platform, but the key differences are that the device tends to be aftermarket and not integrated into the truck. Some lower-cost devices plug into the diagnostics port of the vehicle and are called dongles (see Fig. 2.14). Engine manufactures and some OEMs also sell integrated devices. In terms of functionality, the device connects to the CAN bus, manages the wireless connection, has integrated GPS, but has more input/output ports for integration with analog and digital sensors. These sensors can be used for a variety of functions from trailer door status to power take-off engaged, sensing events that are not available on the CAN bus.

The APIs available for a TCU are connecting to and disconnecting from the IoT layer to send and receive data, very similar to the TCU-related APIs of the connected vehicle platform.

Example operations to connect a vehicle, to send/receive data and instructions:

Action	URL	Description
PATCH	/vehicle/{id}/connect	Confirm SIM voice or data connection of VIN
POST	/vehicle/{id}/event	Send event of VIN with GPS and data
POST	/vehicle/{id}/service	Instruct vehicle to perform action
POST	/device/{id}/command	Send controlling command to device
POST	/vehicle/{id}/sync	Sync vehicle data with cloud

Head Unit

In commercial telematics the head unit is similar to the head unit in the connected vehicle platform, but is designed more for functionality than entertainment. It supports the driver functions for schedule and dispatch, navigation, safety and compliance. The device tends to be aftermarket, is not integrated into the truck and more rugged due to the environment and types of drivers. Tablets and smartphones can also be used as less expensive options or in a less rugged environment.

The APIs are very similar to the head unit APIs described in the Generation 2—infotainment and navigation.

2.4.3   Internet of Things

For commercial telematics, the IoT layer is pretty much the same as described previously for the connected vehicle. The only exception is when the driver uses a smartphone or tablet to run the applications. Then the business process for activations and billing is different. Instead of activating automatically through a TSP or OEM system, a fleet manager buys smartphones and tablets directly and activates them just as a consumer activates a smartphone today.

2.4.4   Mobility Cloud

In the mobility cloud layer, we defined the following two service domains:

1. Fleet and trailer tracking, and
2. Fuel and maintenance.

The capabilities are very similar as described in the consumer-oriented generations.

Fleet and Trailer Tracking

The domain related to fleet and trailer tracking comprises the following services:

Location and status:	One of the first functions implemented in commercial telematics and still one of the most important. It supports a set of the other functions as well from job status in schedule and dispatch to turn-by-turn directions in navigation. What has been added related to the location is the truck and trailer status information, which includes the condition of the truck and trailer, speed, direction, engine status, tire pressure, fuel level, fluid level, power take-off usage etc., all are used by the fleet operator to improve fleet operations.
Breadcrumbs:	A feature that expands on location tracking, breadcrumbs can show a fleet operator a graphical history of where a truck has been over a given period of time. The value comes from confirming to the drivers that they are sticking to the route and aren't taking any side trips, which can cost the fleet manager fuel and wear and tear on the truck.
Geofence:	Geofencing expands on breadcrumbs by defining an area a truck cannot leave or enter without a notification being issued. This is another way to confirm trucks are not taking side trips, but it is also useful for integration with job status so the system can automatically track arrivals and departures from a given area.
Trailer tracking:	Similar to vehicle location and status tracking, but warrants its own services. Trailer tracking is just the location tracking of trailer alone. At warehouses and distribution centers, there are certainly many trailers without trucks, someone actually owns those trailers and sometimes they get lost. So special standalone systems were developed just to track the trailer, since limited battery power can be an issue.
Asset health and tracking:	Along with tracking the trailer, sometimes tracking the assets within the trailer can be important as well. Most common are the refrigeration trailers used for transporting perishable food and making sure the temperature stays within certain limits. Special-asset tracking is sometimes required for high-value assets or assets that can pose a danger if not handled correctly.

Example operations regarding fleet tracking are:

Action	URL	Description
GET	/vehicle/{id}/status	Get vehicle status, e.g. location, speed
GET	/trailer/{id}/status	Get trailer status, e.g. location, battery
GET	/vehicle/{id}/shipment/{id}	Get shipment status, e.g. temperature
GET	/vehicle/{id}/fence	Get polyline geo-fence
POST	/vehicle/{id}/fence	Create polyline geo-fence
DELETE	/vehicle/{id}/fence	Delete polyline geo-fence

Fuel and Maintenance

The domain related to fuel and maintenance comprises the following services:

Remote diagnostics: Similar to connected vehicle monitoring, the diagnostics code from the truck can be very valuable. In the fleet business, vehicle downtime can cause direct loss of revenue, so keeping the fleet healthy is one of the fleet manager's top priorities. OEM engine providers are now providing their own solutions integrated with many of the commercial telematics providers for monitoring the diagnostics codes directly, and even providing the solution for free during the warranty period to help reduce their own warranty charges. In certain cases, getting the correct information to the driver at the correct time can prevent a simple problem from becoming a complex and costly problem. When a system pays for itself during a warranty period, the fleet manager knows there is lots of value.

Predictive maintenance: Similar to the connected vehicle and actually more popular because trucks typically require more maintenance than passenger cars, predictive maintenance goes beyond just monitoring the diagnostics codes. It uses a history of the vehicle data and driver data to better determine when to do maintenance or even predict when maintenance is needed before a problem occurs. The savings associated with remote diagnostics and avoiding downtime can save a fleet owner a lot of money.

Fuel tracking: Fuel is one of the most costly components in operating a fleet, so any savings directly affect the bottom line. Driving scores are one way to optimize driving styles fuel usage. Geofencing helps keep drivers to their assigned routes, but adding the integration of fuel cards directly correlates the fuel purchased with the fuel used. Fuel cards are used by fleets at certain gas stations to avoid using personal cards, an efficient way to manage fuel cost, but theft can be a problem.

Example operations regarding fuel and maintenance are:

Action	URL	Description
GET	`/account/{id}/score`	Get driving scores based on past data
GET	`/vehicle/{id}/diagnostic`	Get diagnostics report of vehicle
POST	`/vehicle/{id}/diagnostic/log`	Log DTC for vehicle
POST	`/vehicle/{id}/diagnostic/predict`	Predict vehicle maintenance

2.4.5 How Are These Services Used by Customers?

Mostly driven by the OEMs, agriculture and construction are special industries within commercial telematics that are examples of how these services are used by customers. Agriculture led by John Deere is not just a commercial telematics solution, but tends to be more of a cognitive vehicle or maybe cognitive tractor in this case. Many people may not be aware but autonomous tractors or precision agriculture have been around longer than all the hype we are currently seeing with autonomous vehicles. So what is precision agriculture? In general, it is the ability to monitor, control and respond to situations in the field. We can find precision agriculture in planting equipment that controls where the tractor is going, when it's planting, how far apart it's planting and how deep it's planting. Originally, planting had basic controls for when a seed was planted and how it was planted. Next they introduced technology to control how each row is being planted, instead of planting all rows at the same time. This optimized the use of seed in irregular shaped areas, when the planter got to the end of the row, and when it turned around to plant the next set of rows. Then guidance was added so we could plot how we wanted a field planted, program that into the tractor and planter, and off we went.

More advanced systems exist now that integrate external data about the environmental conditions of the field, so more or fewer seeds can be planned in certain areas, based on moisture, slope and other conditions. Drones are being used along with other technologies to capture the required information about the field including slope and moisture content. Physical soil sampling is also used. All this is done through location guidance based on GPS and data about the conditions in the field are used to optimize where and how the seeds are planted to improve yields. Spraying equipment also uses similar data from the field and location guidance to optimize where and how much fertilizer is used. John Deere states, "an average of 10% reduction of inputs in planting and fertilizing".[11]

The other solutions from John Deere and Caterpillar focus on job site solutions. These solutions are not just for farms, but for construction sites, mining sites, etc., providing a business solution that manages the operations as well as the typical commercial telematics functions of keeping the fleet of equipment operational.

2.4.6   How Far Is It from 'My Cognitive autoMOBILE Life'?

We are assessing the Generation 2 commercial telematics platform based on the foundation of the six self-enabling themes, (see Sect. 1.6) and illustrate the score in Fig. 2.13:

Self-integrating	3, for the integration of the truck location and job information into the fleet business.
Self-configuring	3, for the IVI that started to replace mechanical functions with digital functions along with support for schedule and dispatch.
Self-learning	2, for the ability to plug intelligence into the truck via smart-phones, but still mostly descriptive analytics.
Self-healing	3, for remote diagnostics that helped the fleet owner directly and the OEMs internally.
Self-driving	1, for getting started with driving assistance functions visualized inside the IVI.
Self-socializing	2, for the truck sharing its location and speed to support live traffic data.

In summary, the both Generation 2 self-enabling assessments Figs. 2.13 and 2.16 are nearly identical except for the improvements in the areas of the self-healing theme. But even self-socializing has the same rating even though it has special attention across trucks and their drivers.

A futuristic example of self-socializing related to commercial telematics is when the CPAs can be used in a more personalize way, see [7] for more details. The key aspect is creating a personalized conversation beyond how social networks do it today.

Social and business networks are combined in today's professional world. Social networks may be used to exchange information even for business applications. For example, commercial fleet drivers have a lot of time during the day while they are driving. Historically, they listen to the radio, listen to their own music, or communicate with workers/friends via a phone or private radio networks. Very popular is still the older technology CB radio, which is limited to short-distance

---

[11]"John Deere Precision Ag Technology" https://www.deere.com/en_US/docs/html/products/precision-ag-technology/precision-ag-technology.html. Accessed: July 26, 2017.

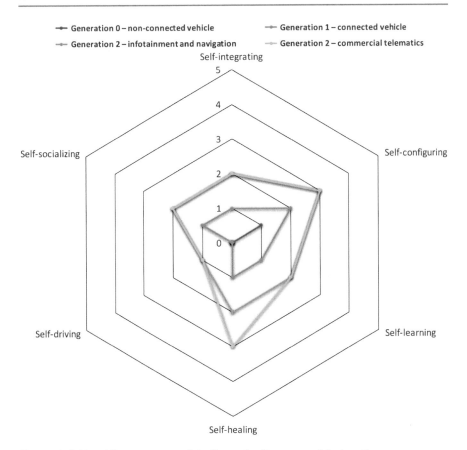

Fig. 2.16 Self-enabling assessment of the Generation 2—commercial telematics

communications on a limited number of channels. The limitation with private radio networks is that the driver is forced to listen to conversations that may not be interesting to him.

Today social networking is built on static or manual associations with topics, people or areas of interest. For example if you are interested in following the South Carolina Gamecock on Facebook, you need to manually search and select to follow them. At best, social networks make a recommendation on areas of interest. So what is a good way to create areas of interest dynamically? Let us define a method to create dynamic topics and subscriptions of those topic based on the user's interest. The method includes analyzing keywords in a person natural language conversation and creating a social networking topic that can later be used as the foundation for a conversation. Other parties in the network can then subscribe to the topic and receive subsequent updates. A person's preferences and usage of topics on the social network may help determine which keywords become topics. A person can later select and rank topics important to them, which also contributes to keyword selection. The method is designed without manual interaction so that it can be used, for example, in vehicles to limit driver distraction and create more personalized content streams, which a driver can listen to. The message can also include context from when the translation occurred including sensor information from the device, i.e. GPS location, accelerometer, etc., which can also be used by subscribing third-party applications.

In [7], a method is proposed for dynamically linking information in a network using NLP on a computer system. The information being contained in a message of a user in the network, where a user interface is provided and adapted for receiving and reproducing the natural language input. A message is defined as a data structure comprising at least the information as well as a set of topics, defined by keywords extracted from the information. A user profile is defined as a data structure comprising at least a set of weighted topics defined by keywords as well as a system of rules used for extracting topics from information.

The method comprises (see Fig. 2.17)

1. recording information from the first user in the network in natural language;
2. translating the recorded information in natural language to a text information;
3. identifying topics in the text information based on topics stored in a predefined database, in particular by comparing the text information to most likely topics stored in the database;
4. combining the identified topics in the message with the set of weighted topics in the profile of the first user;
5. updating the weighting of each of the identified topics in the profile of the first user; and
6. publishing the text information on the computer system.

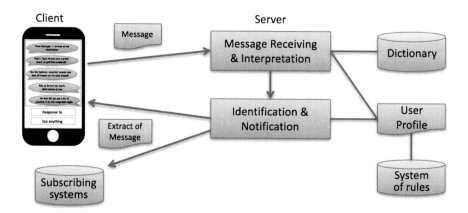

Fig. 2.17 Ad hoc socializing through dynamically linking information in a network using NLP

Fig. 2.18 Smartphone as an
assistant to dynamically
linking information in a
network using NLP

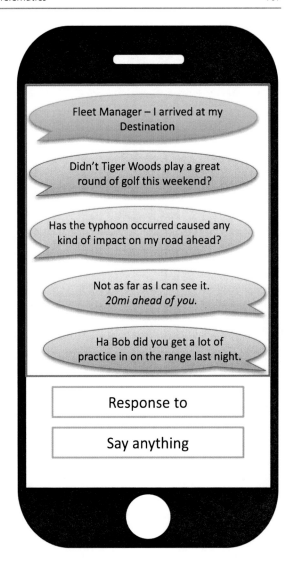

The smartphone can be the cognitive assistant to lead this dynamic socializing
dialog (see Fig. 2.18). The following sample use cases describe how the method
works:

1. The driver presses the button labelled "Say Anything" and says "I would like to
   listen to some sports". The system knows "sports" from the driver's profile,
   but "golf" and certain players are really where the interest is because the
   rules (sport, golf), (golf, tiger woods) are in the system. Topic subscription for
   "golf/Tiger Woods".

2. The driver says "Didn't Tiger Woods play a great round of golf this weekend?". The system selects "Tiger Woods" as a topic for publication.
3. The driver says "I would like to listen to weather in my area". The system knows "weather" from the driver's profile, topic subscription for "weather/area" where the area is defined through the context.
4. The driver says "The weather is really affecting the driving conditions." The system selects the "city and area" based on GPS location and selects city/area as a topic for publication.
5. The driver says "Hey, Bob did you get in a lot of practice on the range last night". The system knows a driving range is associated with "golf", and "Bob" is a good friend. Topic published to "Bob" but not "golf" since the driver only talks to Bob about golf.
6. The driver says "Yes I did, I worked on my putting as well." The system knows it's a response to Bob and publishes the message to him only.
7. The driver says "Hey, Bob I am expecting to get wet on my last few stops". The system knows "Bob" is ranked higher than "weather" and where "wet" is associated with "weather" in the system of rules. Topic published to "Bob" but not "weather".
8. The driver says "Same here, let us keep our umbrella handy." The system know it's a response to Bob and publishes the message to him only.
9. The driver says "Dispatch, tell all my customers I am making good time and I am on schedule for the remainder of the day." The system knows dispatch is highly ranked for work. Topic published to "dispatch", with location, time, and directions and publish arrival alerts to all "customers".
10. The driver says "I was making great time today until I had issues with a flat tire". The system doesn't recognize any defined topics in the dictionary, but scoring starts on keywords like "issue" "flat tire" while analyzing past communications.

2.4.7   How Relevant Is It Across Industries?

Commercial telematics is starting to overlap with solutions and companies in the consumer space. Uber, a leader in the ridesharing market, is using its platform and concepts to solve problems in other industries as well. For trucking, one of the biggest problems is returning from a delivery with an empty trailer. Today, companies known as freight brokers exist that can be contacted and can help with assigning shippers to carriers. The process to is not very efficient say the least. Enter Uber Freight[12] Using the same concepts as ridesharing, owner-operators can now simply log on to an mobile application and quickly see available loads in their area, just like in ridesharing where we can quickly see if rides are available in the area and how long it will take. The value comes from quickly being able to find a load, and improvements in terms of payment are a great advantage for the driver as well.

---

[12]"Uber Freight" http://freight.uber.com Accessed: July 26, 2017.

Another example of industry overlap is UberEATS, where one of the core functions in commercial telematics, schedule and dispatch, is replaced with a dynamic fleet of drivers and the Uber delivery sharing application. The concept is simple, instead of restaurants having their own delivery service, UberEATS[13] signs up drivers and restaurants, so when someone wants a delivery, the driver signs up to bring the food from the restaurant to the consumer location. These are advantages for the restaurants in not having the overhead for their own delivery service. The value for the consumer is an application that shows how long the order will take before arrival, with the same type of accuracy we see in Uber's consumer application. Another example of traditional industry and functional lines being blurred by new companies and participants.

In summary, commercial telematics has tremendous value and that is probably why it is so successful in the industry today. The previous examples show how process optimization, cost reduction, safety, and improving the customer experience are all factors in the value provided by these solutions. It would be nice if this was the case for all industries, but the examples are at least templates for other industries to follow.

2.5   Generation 3: Cognitive Vehicle (Today)

What makes up a *cognitive vehicle* today?

At the very start of the introduction (see page 3) we simply summarized the cognitive vehicle as not only a physical but a personalized mobility experience managed by a CPA. More detailed, we used the following six self-enabling themes as a foundation for defining the maturity of the cognitive vehicle more precisely in Sect. 1.6:

- Self-integrating,
- Self-configuring,
- Self-learning,
- Self-healing,
- Self-driving, and
- Self-socializing.

And in Sect. 1.5 we defined the context of the personalized mobility experience. But we will not achieve the described level of personalized mobility experience with the capabilities in production today. Too much is changing with this generation, therefore, we are separating this Generation 3 into two evolving solution architectures, one which is feasible today and one that will come around 2022.

---

[13]"UberEATS" http://www.ubereats.com Accessed: July 26, 2017.

2.5.1   Solution Architecture

The two new key capabilities to enter into the cognitive vehicle are the cognitive personal assistant in the mobility cloud and the instrument cluster in the vehicle. Each enhance many capabilities by personalizing or even creating new interactions with a focus on a more personalized customer experience.

The first examples that are available today are infotainment preferences to adapt the heating, ventilation, air conditioning, destination planning, smartphone synchronization, and synchronization with home and work. But we are seeing NLP increasingly getting into the vehicle to provide a natural voice interface to these new capabilities with a focus on more complex personalization functions and on simplifying the HMI.[14] The personalized part comes from storing the personal settings, preferences of the driver so the technology can tailor the information and also connect to outside data in a merchant ecosystem.

The natural voice-based dialog is going beyond most systems today that can only answer a single question at a time with no relationship between one question and the next. A dialog knows the context of the previous question and can be used for more complicated tasks like ordering food. Imagine multiple passengers sitting in a vehicle while driving and needing to be synced for a food order accounting for all of their special wishes.

This new integration of merchants builds on the functions of Generation 2, infotainment and navigation, where the user has the ability to interact with merchants in the form of advertising or ordering transactions. Integration into OEM loyalty programs or customer loyalty programs makes this a more personalized experience as well. This is how we build towards the next Generation 4, a cognitive life where the home, vehicle and office come together in a new type of customer experience.

Further applications are

- analyzing driving behaviors based on braking, acceleration, and steering,
- monitoring the environment, i.e. traffic, pedestrians, buildings and even objects

all analyzed using machine learning techniques to output driver alertness, driver rating and route planning.

Figure 2.19 shows the solution architecture with the service domains and capabilities required to deliver them.

---

[14]"Your Car Could Become Virtual Personal Assistant—and Even Know When You've Had a Bad Day at the Office" https://media.ford.com/content/fordmedia/fna/us/en/news/2017/02/21/virtual-personal-assistant-car.html. Accessed: July 26, 2017.

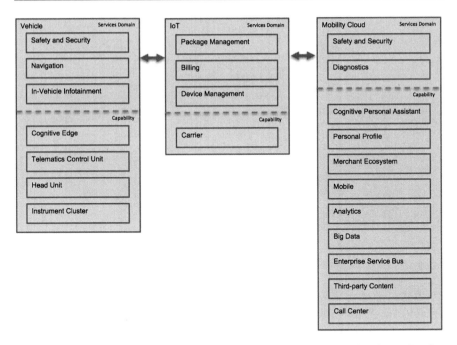

Fig. 2.19 The solution architecture of service domains and capabilities for Generation 3—cognitive vehicle (today)

2.5.2   Vehicle

In the vehicle layer, we enhance the service domain navigation by adding dynamic POIs and adding two new capabilities

1. Cognitive edge as the built-in vehicle capability of the cognitive personal assistant and
2. Instrument cluster to display information with a digital readout behind the steering wheel.

Navigation

Building on the POI definition, we introduced static specific places in the service domain navigation in Generation 2 (see p. 78). Now more dynamic types of POI are starting to emerge in the industry. Most are related to safety and updating incidents in near real time on highly accurate navigation systems or ADAS.

Both standards bodies, the SAE[15] and the European Telecommunications Standards Institute[16] are working on standards to support the data exchange of

---

[15]"SAE International" http://www.sae.org. Accessed: July 26, 2017.

[16]"ETSI—European Telecommunications Standards Institute" http://www.etsi.org. Accessed: July 26, 2017.

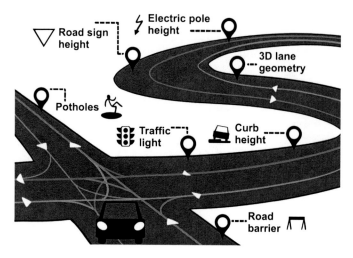

Fig. 2.20 Vehicles are sharing dynamic types of POIs, which are mostly related to safety

information between vehicles. The SAE standard is 'J2735—Dedicated Short Range Communications' and contains data related to the vehicle's size, position, speed, heading, acceleration, route history, wipers, braking system status, and more. From this type of data we can derive the type of scenarios and new POIs that become available to vehicles, specifically when vehicles are sharing information. Incidents such as the end of a traffic jam, recent accidents, sudden stops, potholes, hydroplaning, and more can be shared now and in near real time to improve safety (see Fig. 2.20). The changes are becoming very dynamic. Nowadays, HERE makes around 2.7 million changes to its global map database every day.[17]

Example operations for custom location extension inside a navigation system are:

Action	URL	Description
POST	/map/layer	Upload POIs to map layers
GET	/map/layer	List geometry layers
GET	/map/layer/{id}	Get geometries in map layer
PATCH	/map/layer/{id}	Modify geometries in map layer
DELETE	/map/layer/{id}	Delete geometry layer

More comprehensive is the example of HERE REST APIs & Platform Extensions.[18]

---

[17]"Autonomous cars can only understand the real world through a map" http://360.here.com/2015/04/16/autonomous-cars-can-understand-real-world-map/. Accessed: July 26, 2017.

[18]"REST APIs from HERE Maps—HERE Developer" http://developer.here.com/develop/rest-apis. Accessed: July 26, 2017.

Cognitive Edge

With the use of smartphones today, most people don't think about connectivity. They assume their smartphone is connected all the time and outside of traveling on a plane don't even think about how it operates in disconnected mode. But for safety reasons, the AUTOmobile is unique with the requirement that it must function all the time in a disconnected state and being connected is only an option that has the ability to provide better services and data from the cloud. This is how Generation 1 and 2 have evolved. But especially with the introduction of IVI and the enhancements achieved by integrating the smartphone, the user experienced a new level of personalization and is used to leveraging nearly unlimited resources and possibilities. Hence, a concept inside the cognitive vehicle is needed that meets the safety requirements but provides the personalized smartphone experience while driving. Remember the challenge to model the CPA we illustrated in Fig. 1.2.

For the vehicle layer, we introduced the cognitive edge, which is a core set of capabilities that will drive the cognitive vehicle of the future.

> The concept of how to manage the edge with the cloud in a connected and disconnected fashion providing AI services is the core of the *cognitive edge*.

What data to send, how much to send, when to send it, all configurations that need to be managed and in sync with the cloud. As we discussed in defining big data, just because we have a lot of data, doesn't mean we need to collect it all. These design decisions start at the edge with basic designs around filtering, rules, and configuration that control which data are even sent.

What makes it more like an application is the ability to manage where information and events go, one configuration may be that all traffic events are displayed on the head unit, where preferred POIs are on the head-up display. With the complexity of human interaction changing, simple point applications on a head unit are no longer valid. Information may be presented as audio, blinking lights on the dashboard, points on a head-up display, haptic signals in the steering wheel, etc.— clearly more complexity to manage.

The cognitive edge is the built-in vehicle capability of the CPA leveraging STT, TTS, NLP and further AI technologies to function smoothly with the CPA in the cloud even when the vehicle gets disconnected from the Internet. What language models are synchronized with the cloud, what voice functions work in disconnected mode, again all configurations that may vary by OEM.

The complexity now lies in what to do with personalized data and how it is synchronized with the cloud. To drive a more personalized experience, a driver's preferred POIs, preferred merchants, vehicle settings, driving characteristics etc. all need to be securely managed and synchronized with the cloud. Not only will information be important at the edge and in the cloud, but the analytics we do on it and each place will be as well. Some types of analytics that require large amounts of data and data processing will always be done in the cloud, like determining travel patterns or driver behavior. Driver behavior can also be an example where edge analytics can be used on real-time data to determine certain behaviors like drinking and driving or begin drowsy.

The HMI cannot simply tell the consumer that the vehicle connection is lost and the service used a minute ago cannot be used or that its functionality is limited. Smarter responses need to come through the cognitive edge like the CPA says "I need more time to answer your question."

In the future, these trade-offs between edge and cloud will differentiate brands and the type of services and functionality they offer. Cost and processing power are the determining factors in edge analytics or any edge-type processing, both of which have improved in recent years, but it is still not like having a fully functioning smartphone or tablet in the vehicle. The APIs are typically the embedded or edge versions of their cloud counterparts, for CPA, personal profile, and analytics. The main difference is that the data used are synchronized with the cloud.

Example operations regarding cognitive edge are:

Action	URL	Description
POST	`/vehicle/{id}/sync`	Sync vehicle data with cloud
GET	`/nlp/tts/voice`	List available voices, e.g. language
GET	`/nlp/tts/voice/{id}`	Get information about voice by ID
POST	`/vehicle/{id}/nlp/{id}/sync`	Sync language model with cloud
POST	`/vehicle/{id}/account/{id}/sync`	Sync user profile with cloud
POST	`/vehicle/{id}/service`	Instruct vehicle to perform action

It is important to differentiate the following two APIs:

Action	URL	Description
POST	`/account/{id}/device/{id}/connect`	Connect a user device to head unit
POST	`/vehicle/{id}/account/{id}/sync`	Synchronize user profile with cloud

In the case of the first API, a user-owned device like a smartphone is connecting to a vehicle device such as a head unit. In the case of the second API, the cognitive edge built into the vehicle is in the active process of synchronizing with the user profile data in the cloud.

Instrument Cluster

> Traditionally the instrument cluster is a set of instrumentation directly behind the steering wheel displaying information like speed, fuel level, temperature, tire pressure, and more.

Recently, OEMs have started to display other information from the IVI such as the radio station and artist that's playing or maybe parts of the navigation system's turn-by-turn directions. Head-up displays are also an extension of the instrument cluster showing similar information, but directly reflected off the windshield. In the future, information will become more digital and configurable, so the lines will blur on what is the head-up display, instrument cluster or IVI. Look at startup Carrobot,[19] a company focusing on a head-up display with 4G, WiFi, Bluetooth, Full Voice Control, Entertainment, AI and "The Smartest HUD device in the World". A great example of companies already blurring the lines between the head-up display, instrument cluster and the head unit. It is also a product that incorporates capabilities of a CPA, so it is not just a display, but uses voice interactions as well. In the past, aftermarket devices have set the trends for automakers, Garmin navigation is one example and Bluetooth hands-free kits another, so these aftermarket-type head-up displays can certainly give us a perspective on where OEMs should go. Another example is Panasonic as a supplier of both head units and head-up displays[20] The company is certainly in a position to integrate these types of solutions as well.

Example operations regarding instrument cluster are:

Action	URL	Description
POST	/device/{id}/display/speed	Display speedometer on instrument
POST	/device/{id}/display/fuel	Display fuel level on instrument
POST	/device/{id}/display/economy	Display fuel economy on instrument
POST	/device/{id}/display/travel	Display travel direction on instrument
POST	/device/{id}/display/economy	Display outdoor temperature

2.5.3   Internet of Things

The IoT layer continuously improved each generation in multiple areas. Wireless connectivity speed offered by the carriers improves every few years, with 2G, 3G,

---

[19]"Carrobot" http://carrobot.com. Accessed: July 26, 2017.

[20]"Heads Up Display Solutions—Panasonic Automotive Solutions" http://business.panasonic.com/solutions-automotivesolutions-ecockpit-headupdisplay. Accessed: July 26, 2017.

4G, and even 5G in the future. In concert with declining costs, this results in more and more of the technology making it into vehicles. The carriers started sharing data plans with customers to reduce costs and integrating that with Wi-Fi hotspots in the vehicles as well.

Most enhancements in the IoT layer are in the service domain 'Device management'.

Device Management

Device management is improving with more OTA updates of software in the vehicle. Both on the head unit or even other ECUs which can save the OEMs from bringing the vehicle into the dealers if problems are related to software only, which is becoming more and more the case. An early lesson learned: Do not put a device in the field that doesn't have a failsafe mode for communication and cannot be updated over the air. Initially not a big problem for OEMs, since most people are accustomed to going to the dealer, but if the product is a refrigerator, for example, that is a bit different.

OTA does have its complexities as devices or systems matures. Take updating the ECUs in the vehicle, Firstly, not all ECUs even have this capability. Secondly, the version management and dependencies required by the OEM to perform the function isn't as straightforward as consumers would expect. In addition, there is the complexity at the edge to manage making all the changes and also be able to recover if a problem occurs in the process. This usually involves different providers and integration with an OEM, so really it's not that easy to do in complex situations, but the basics of updating the main communication device is handled by most OTA providers in the market. In Chap. 4, we will discuss a different type of OTA that's more like the app store for the smartphone, but changes more the characteristics of the vehicle.

Example operations regarding OTA in device management are:

Action	URL	Description
GET	`/deployment`	List deployments
GET	`/deployment/{id}`	Get a deployment
PATCH	`/deployment/{id}`	Update a deployment
POST	`/deployment/{id}/config`	Create a config
POST	`/deployment/{id}/package`	Create a package
GET	`/device/{id}/software`	Get software version
PATCH	`/device/{id}/software`	Update device software

2.5.4   Mobility Cloud

In the mobility cloud layer, we added three new capabilities,

1. the core is the CPA that is leading a contextual conversation,
2. a personal profile to collect, store and manage the data related to a person, and
3. a merchant ecosystem to enable advertising and interactions between merchants and the connected vehicle.

In addition, the capability analytics from the previous Generation 2 will be enhanced with more functions.

Cognitive Personal Assistant

The *cognitive personal assistant* (CPA) leads a contextual conversation to interact with a human touch, collects and analyzes information to get more insights, and connects to everything to link capabilities for personal advantages.

Figure 2.21 shows the CPA with the core AI from contextual conversation leveraging big data and ecosystems. A challenging goal is that a CPA communicates

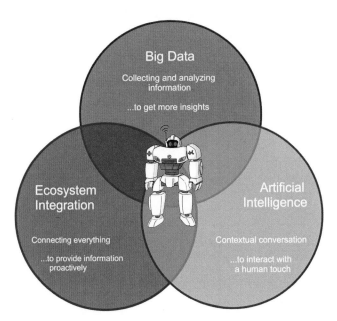

Fig. 2.21 The CPA with contextual conversation capabilities as the core intelligence leverages big data and ecosystems

with the same fluency and articulation as humans in the requested context. In the NLP Sect. 1.2.3, we introduced examples that are feasible today, i.e. the CPA can mimic realistic-sounding human voices and can lead linear dialogs and with some limitations also some specific non-linear dialogs, which is more than a question-answering system. It is getting the core HMI for the next platform generations. We refer to [2] for some early design thinking in this area. In addition, a CPA leverages the big data and analytics capabilities we introduced previously in Generation 2. The CPA also leverages the ecosystem integration capabilities of the ESB and third-party content we already introduced in Generation 2.

Example capabilities are: The CPA

- uses speech to interact with the driver and passengers of the car.
- is able to participate in complex conversations.
- explains things in an appropriate way for different persons.
- reacts to emotions of the driver and passengers.
- gives individual recommendations based on behavior.
- proactively informs about upcoming events.
- incorporates multiple data sources to get more insights on the persons.
- steers the car.

The APIs for a CPA will only focus on the speech and basic conversation aspects. The intelligence and ecosystem APIs will be discussed in a later generation. In the vehicle, a driver either will push a button or in future the vehicle will always be listening for keywords to start a specific action. This is how Alexa works in home automation today instead of pushing a button on a phone to activate Siri, for example. Once the CPA is initiated, the speech APIs are used to process our voice into text and text back into voice. The NLP (see Sect. 1.2.3) is divided into many different APIs to deeply analyze the input text. For example, one API is used to categorize the input into a classification of levels based on the understanding of the text. Level 1 is a high-level category like finance, where level 2 goes deeper into business or personal finance. The next level separates lending versus spending and the final level a credit card or checkbook. This type of classification helps to narrow the understanding of the text to better analysis it with other APIs. Recognizing concepts and relations between entities are diving deeper into the text analysis. For example, analysis of a text about deep learning would likely return the concept AI, even if the term AI is not explicitly mentioned in the input. Important for the CPA is to also analyze the tone in its own customer engagement during the conversations.

Example operations for a CPA are:

Action	URL	Description
GET	/nlp/tts/voice	List available voices, e.g. language, gender
GET	/nlp/tts/voice/{id}	Get information about voice ID
POST	/nlp/tts/synthesize	Synthesizes text-to-speech
GET	/nlp/stt/model	List available models, e.g. en-US broadband
GET	/nlp/stt/model/{id}	Get information about model, e.g. min. sample rate
POST	/nlp/stt/recognize	Send audio to get transcription results
POST	/nlp/analyze	Analyze features of natural language content
POST	/nlp/categorize	Categorize input into a classification
POST	/nlp/concept	Recognize concepts that are related to the input
POST	/nlp/emotion	Detect emotion conveyed by the entire input
POST	/nlp/entities	Identify entities, e.g. people, cities
POST	/nlp/relation	Recognize related entities, identify relation type
POST	/nlp/tone	Analyze general tone

More comprehensive are examples from the IBM Watson APIs.[21]

Personal Profile

> A *personal profile* has the ability to collect, store and manage the data related to a person.

Today, vehicles have the concept of a driver profile, i.e. driver 1 and driver 2, but as we look at the cognitive vehicle the concept of a personal profile is really needed to customize the personal experience. Firstly, it requires the idea of identifying the driver because multiple drivers may use the same car key. The industry is currently looking at multiple ways to identify the driver, either by built-in video, voice, other biometrics like weight or by the smartphone the driver carries. With the increase of cameras and the multiple uses for them in the vehicle, facial recognition is one way to easily identify the driver. This may touch on privacy issues, but for identification purposes plus the ability to monitor a driver for drowsiness, eyes on the road, and emotions, cameras are a good solution.

Speech recognition can also be used for most of these same functions apart from eyes on the road. OEMs will probably implement one or both capabilities in the near future. IBM and others can analyze the tone of the driver's voice or the types of words used to gain a better understanding of the driver's state. Emotions and

---

[21]"Watson services—IBM Watson Developer Cloud" http://www.ibm.com/watson/developercloud/services-catalog.html. Accessed: July 26, 2017.

the current state of the driver are becoming essential tools in personalization, for example if the driver is perceived to be tired, maybe suggesting a break or coffee is a safe recommendation to make. But the e-motion is elementary not to get a driver's angry reaction: "I'm not tired!" That is not achievable today.

In the commercial trucking industry, safety and driver monitoring are not facing the same privacy debate, but the value of identifying a problem with the driver is good for both the owner and operator of the vehicle. Once the driver is identified, the driver can create and manage his profile. A profile can include radio settings, seat settings, IVI settings, those typically associated with the car key, but will need to expand to other information including social media such as Facebook or Twitter. Other information can also include merchant preferences, driving preferences, and any other data that can help to personalize the experience.

Creating and managing personal profiles are not new, Google, for example, is known for building a personal profile based on a user's search history, mobile applications, and other Internet activities where the user leaves a digital footprint. The automotive industry is a bit more concerned about privacy, but can certainly adopt other generally accepted practices as it moves into more personalized services, which are essential at least to get closer to the creepy line we drew in Fig. 1.5. Of course, as the vehicle contains and manages more personal information, different types of security and protection of the data will be needed. In Chap. 4, we will explore further technology options that can be used to better secure this information.

The APIs around the personal profile are the first example of some of the intelligence and core capabilities provided internally by the CPA. As the systems collects voice or video inputs, the driver identification and traits can be captured and analyzed in real time, and used to build a robust personal profile over time. That is very different from the account related APIs we introduced with the IVI in Generation 2. Example operations for a personal profile are:

Action	URL	Description
GET	/device/{id}/service	List of services discovered for device
GET	/device/{id}/data	Get a device captured data
POST	/account/identify	Send devices data to identify account
GET	/account/{id}/voice	Get a user's voice characteristics
GET	/account/{id}/visual	Get a user's visual characteristics
GET	/account/{id}/trait	Get a user's trait
GET	/account/{id}/preferences	Get a user's preferences
GET	/account/{id}/device	List of a user's devices, e.g. Bluetooth
GET	/account/{id}/device/{id}/service	List of services discovered for device
GET	/account/{id}/vehicle	List of a user's vehicles

Merchant Ecosystem

> A *merchant ecosystem* is similar to an ecosystem of third-party content providers, but the main difference is that the content of the merchant is provided by them as a service in the form of an advertisement, coupons or an opportunity to purchase.

For advertising and coupons, a campaign management system is required to manage the content and distribution of the content to the vehicle. The content is usually offered as service directly by the merchant and does not need to be integrated through an ESB like the third-party content. For purchases, a payment system is required to exchange money with the user and the merchant. But the model of how and when to pay is still evolving, it can be the TSP, the OEM, or even the merchant. More will be discussed about the related business models in the next Chap. 3.

In this Generation 3 of today, the concept of pushing advertisements to a vehicle is not very appealing to OEMs or consumers, so concepts based on location or upon request from the user are better scenarios. The ecosystem is the ability to bring multiple merchants together, allowing the OEMs and customers to select merchants based on their preferences.

The industry and CPAs are starting to see may different types of ecosystems develop around products.

- Amazon, for example, has skills and merchants it keeps adding to their Echo product driven by Alexa, their CPA foundation. Drivers in the future will be able to access Alexa, but the integration into the vehicle may only be limited to location and not vehicle data, which is really needed to enhance an experience that might change in the future as well.
- Google has a similar ecosystem around their Google Assistant, but with OEMs starting to put the Android operating system into head units, a future Google Assistant will have access to the vehicle information to create a much better user experience. One issue OEMs have with Google is giving up part of their customer experience and driver information to Google, clearly some OEMs will take this route, others will not.

We will talk about this more in relation to a cognitive life, but now introduce at least the core components of a merchant ecosystem. For the APIs, first we need the management components of adding, updating, and removing merchants. Once a merchant is set up and its specific type of content is available, a CPA can either interact with a merchant based on events such as location or push content based on events as well.

Example operations for a merchant ecosystem are:

Action	URL	Description
GET	/merchant	List of merchants
POST	/merchant	Add a merchant to ecosystem
GET	/merchant/{id}	Get information about merchant
PATCH	/merchant/{id}	Update merchant information
DELETE	/merchant/{id}	Delete a merchant from ecosystem
GET	/merchant/{id}/taxonomy	Get category classification used by merchant
GET	/merchant/{id}/search	Search items on merchant catalog
GET	/merchant/{id}/review/{id}	Get item reviews written by users
GET	/merchant/{id}/service	List of services offered by merchant
GET	/merchant/{id}/metric	Data metrics, e.g. earnings, clicks
GET	/merchant/{id}/trend	What are the merchant's top sellers right now
GET	/account/{id}/preferences	Get a user's preferences
GET	/account/{id}/channel	List user's channels for ad exchange
POST	/account/{id}/channel/{id}	Push ad/interaction to user's channel
POST	/payment	Make a payment
POST	/payment/{id}	Payment details, e.g. approved/declined
POST	/payment/{id}/void	Void a payment transaction
POST	/payment/{id}/refund	Refund a processed payment

Analytics

We introduced analytics earlier in Generation 2, but now we expand to predictive analytics and also start using prescriptive analytics (see Fig. 1.8) in combination with spatial analysis, all as ways a CPA can start to understand situations and even learn from them.

Trend analysis and predictive analysis are popular when looking at vehicle data for warranty and diagnostics analysis. Trend analysis in warranty or manufacturing areas look for patterns in the data or trends in what is happening. Upward trends in warranty claims may lead to recalls. Predictive analysis looks for patterns in the data as well, but those that can be associated with known problems lead to the ability to predict a failure before it happens.

What is really becoming differentiating is combining all these data sources,

- the vehicle data,
- the personal profile data, and
- data from the merchant ecosystem

that create a new direction of learning for the cognitive vehicle.

The aspect of looking at these data sources in real time also differentiates this type of analytics and models from the traditional models used in cases of warranty analysis, which are more batch or background process-type of analytics. A CPA can then use an API in real time for a recommendation to get fuel or the driver can be prompted based on the real-time events and analytics.

Example operations regarding analytics are:

Action	URL	Description
GET	/analytics/stream	List data streams
POST	/analytics/stream	Create new data stream
GET	/analytics/stream/{id}	Get data stream description
DELETE	/analytics/stream/{id}	Delete data stream by ID
GET	/analytics/stream/{id}/tag	List tags for a data stream
POST	/analytics/stream/{id}/tag	Add tags to data stream
PATCH	/analytics/stream/{id}/tag/{id}	Update tags for a data stream
GET	/analytics/stream/{id}/record	Get data record from stream
PATCH	/analytics/stream/{id}/record	Write data record into stream
GET	/analytics/forecast	Get all forecast jobs
POST	/analytics/forecast	Start forecast job
GET	/analytics/forecast/{id}/status	Get forecast job status
GET	/analytics/forecast/{id}	Get forecast result
DELETE	/analytics/forecast/{id}	Stop forecast job
POST	/vehicle/{id}/diagnostic/predict	Predict vehicle maintenance

2.5.5   How Are These Services Used by Customers?

The introduction of a CPA is the new key capability of a cognitive vehicle that can be achieved today. However, voice recognition in the vehicle has actually been around for a long time. Vehicles implemented voice recognition over 15 years ago with basic commands for the IVI, so scenarios like

- "turn on the AC"
- or "play the Beatles"

were mainly introduced to reduce driver distraction and provide a simpler interface to the system. The most useful scenarios were related to the navigation system.

Instead of entering an address manually, the driver could say the address out loud and save on time and reduce complexity.

Voice recognition in a vehicle is more difficult than people may think. A combination of background noise and the type and location of the microphone built into the vehicle make it difficult to perform the basic STT operation. This was further complicated by limited grammar and commands and having to know which commands to say and when. These early system were really never quite accepted and plagued with customer satisfaction issues. However, we did see the core components of a CPA put into place on which to build a foundation.

As smartphone became more integrated into the vehicle more recently, not only did we get Siri as a CPA, but access to information through the Internet, so scenarios like

• "what is the weather today"
• or "who won the South Carolina game last night"

became possible. Also rich components around NLP were introduced beamed-in from the cloud, leveraging the cognitive edge. The user was thus not limited to specific commands mostly because of the limited device capabilities. The driver could talk more naturally and questions could be understood. Siri of course is great for information on the Internet or data in applications on user's smartphone, but data integration and information from the vehicle is still lacking.

Nowadays, we start to see the integration of vehicle data with more personalized information about the persons or merchants. Scenarios like

• "your fuel is low, would you like to stop at your favorite gas station?"

require real-time data from a vehicle, analyzing personal preferences and predicting it's time to stop and get gas. A simple response to receive in a vehicle, but requires more complex capabilities in the background to make it happen, but clearly a better user experience. This type of scenario brings together many of the components in a cognitive vehicle, from event management, knowing when the gas is low, to the analytics for predicting the best time to stop. The personal profile is also used from managing personal preferences, tied to a merchant ecosystem that allows connecting with personally preferred providers. The CPA plays an important role in all these scenarios as NLP and enriched dialog create a much better user experience (see Fig. 2.22).

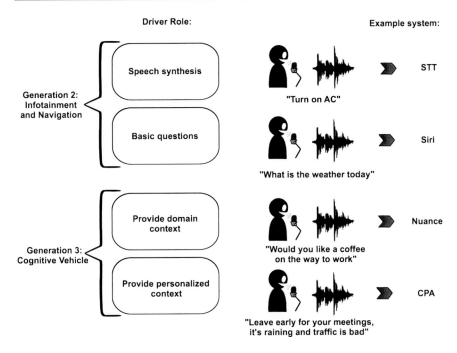

Fig. 2.22 Evolution of the conversational capabilities with consumers through the platform generations

2.5.6   How Far Is It from 'My Cognitive autoMOBILE Life'?

We assess the Generation 3 platform (today) by the foundation of the six self-enabling themes (see Sect. 1.6) and illustrate the score in Fig. 2.23:

Self-integrating   3, for content from merchants, the connection to an ecosystem, and further integration of smartphone capabilities.

Self-configuring   stays the same at 3, because the personal settings of today are still not a real personal profile as defined based on the person's identity.

Self-healing   3, as remote diagnostics improved and the ability to even predict problems helped the customer directly and the OEMs internally for the overall service experience before and after customers are at the dealer.

Self-learning   3, for expansion to predictive and prescriptive analytics capabilities as ways for a CPA to be able to start understanding situations and even learning from them.

Self-driving   2, for partial automation through ADAS and controlled scenarios and spaces for SAE Level 2, e.g. traffic jam assistant.

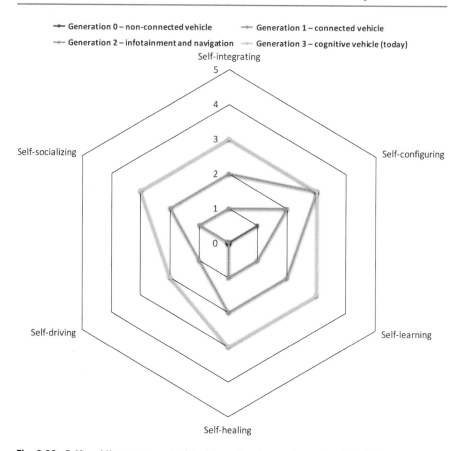

Fig. 2.23 Self-enabling assessment of the Generation 3—cognitive vehicle (today)

Self-socializing   3, for starting to see the vehicle sharing more information based on context and as the smartphone integration plays a more important role in the vehicle to improve customer experience, e.g. by leveraging the merchant ecosystem.

2.5.7   How Relevant Is It Across Industries?

The lesson learned for other industries as we have seen the automotive industry mature from a connected vehicle to a cognitive vehicle can be summarized as a personalized experience, event management, and leveraging an ecosystems.

Let us start with a personalized experience, as companies start to connect products and make them more intelligent in this new world of the IoT. Getting connected is certainly the first step but what we connect with also helps drive the personalized experience. Take for example a connected appliance. It is good that we can preheat our oven before we get home or be alerted if our children are playing with appliances they shouldn't be, but what about connecting to recipes or dinner planning sites. What was command and control type function is now an ecosystem that helps the consumer plan dinner and pair it with the right wines.

Understanding the HMI and whether a CPA can also assist in this type of experience would be interesting to investigate. Sometimes a CPA isn't the answer, maybe a smartphone application that explains the features of our oven better and integrates that with the recipes is more valuable.

Another example may be a connected camera that links to Facebook for social media or Google photos for picture sharing. The steps to simply share a photo on Facebook aren't as easy compared to doing it directly on an iPhone, for example. Or even the ability to capture GPS so the photos can later be tagged with location. These types of improvements in customer experience can really make a difference when products are becoming connected.

Back to appliances. Nest Labs did a great job of personalizing the basic concept of home climate control and integrating it with energy management, but how about integrating it with a service plan so the quarterly maintenance can be scheduled or the owner can even be notified when parameters are out of specification. While renting a home, the renters may neglect to change the filters, which may or may not have led to a problem. But this is an example of a parameter that, if monitored, could have been detected. Predictive maintenance and machine monitoring are the most popular scenarios for connecting products, with advantages for both the owner and manufacturer. But one of the most fundamental capabilities we see as we move into cognitive products is learning and how learning improves and creates a much better personalized experience. This is what Nest Labs has achieved by learning the user's behavior, when he gets up during the week, when he comes home, how that changes on the weekends, and then using that information to set up the climate control system and optimize the energy savings.

Alexa is a CPA, but what makes it really valuable is it is connected to the ecosystem of Amazon. We will talk more about the CPA later, but if we look at the product itself, it's nothing really innovative. Basic CPA functions, voice-activated technology have been around for a long time, but connecting to the ecosystem of Amazon is what really drives the value. Amazon is also expanding on this ecosystem allowing third parties to connect their products, like appliances and even vehicles. We will expand on this more in cognitive life, but for now we stress the integration of an ecosystem and the value it brings.

OEMs have a lot to gain from connecting devices as well. We talked before about warranty savings and other values, but as connected moves to cognitive, the value to engineers and their processes around design and development can change dramatically.

First let us start with the data, in some cases, this is the first time engineers get real data from the field. Sure they do testing and trials, but nothing is better than real customer experiences and data. Feeding this into engineering processes and analyses can help engineers get a better understanding of the product behavior in real-world scenarios. Not only are data available for the first time, but this may be the first time engineering or marketing knows who their customers are.

In the appliance industry, it is common in the United States that products are sold through Lowe's and Home Depot and not directly by the manufacturer, so getting the customer's name is difficult to almost impossible. The fraction of people that will fill out the OEM warranty card is less than 10%. Now engineering and marketing can start to look at customers' behaviors in real time, not through marketing studies or trials. Using this type of information can be fed directly back into product design for feature improvements or removals, what leads to design thinking (see Sect. 1.4).

The insurance industry is actually quite mature in the areas of PAYD and Pay Where You Go insurance. It is a role model for how products change based on learning behavior. The fundamentals behind PAYD insurance are simple. Reward drivers with good driving behavior or simply charge them only when they use their vehicle and not based on a standard monthly rate. Learning driving behavior has a direct overlap with the automotive industry, but the concepts are the same. Measure hard braking, speeding, acceleration, cornering, etc. and the insurance company can understand how a customer drives and how this can ultimately be correlated to their insurance rate. 'Pay where you go' is similar except it is based on the areas the customer drives in. It considers whether it is in a city or a high-accident area, even to the point where suggested routes can be altered to avoid high-priced areas. These types of recommendations while the customer is driving can lead to savings on insurance. Some people are concerned about this type of monitoring, but for example, some vehicles sit more at the airport than they are driven by business travelers and this is just the type of solution for them.

Probably the most famous and best-understood type of learning is in advertising and used by Google, Facebook and others. Google monitors our behavior and searches on the Internet, uses and sells that to others so merchants can engage in targeted marketing. For example, searching for new golf clubs, we would search the OEM websites, search for reviews from industry sites, and sure enough the next time we are on Facebook, read Google news or visit Amazon's website, we get advertisements for golf clubs. This may be scary for some with regards to protecting their privacy, but is pretty simple nowadays for a personalized experience while publicly browsing the Internet.

2.6   Generation 3: Cognitive Vehicle (2022)

What will be the customer expectation in the year 2022?

Customers increasingly expect a personalized experience, demand it in a shorter timeframe than before and want it delivered through their channel of choice. The AUTOmobiles has already been around for 130 years. But the rise of the smartphone dates back just 13 years. We expect the majority of the customers, the digital natives growing up with the Internet, to adopt a new smartphone generation every 1.3 years at the latest to make decisions around their personalized mobility experience in less than 1.3 s. That is the challenge we are facing with this Generation 3 in developing the cognitive vehicle that cannot be released later than 2022.

Now the year 2022 is not far away from an automotive perspective, but that year is seen as a milestone in the automotive industry for maturing some cognitive vehicle concepts like we introduced in Generation 3 that are feasible today, see for example the forecast [4].

Let us take one of the examples we used earlier to introduce the higher maturity of the cognitive capabilities before we go into the definition of the solution architecture. The scenario of getting into the vehicle in the morning and being asked by the CPA

- "would you like to order your normal cup of coffee on the way to work, and the traffic is normal today and will not affect your arrival time at the office".

The first capability needed by a CPA is big data, collecting all the data from the vehicle like GPS, stops, acceleration, hard braking, etc. or data related to where we are going and how we are getting there. If a history of this data is captured, stored in a personal profile, and then basic spatial analytics has been performed, the CPA can learn the trends and patterns for drivers, and then be able to predict where they might be going anytime they get into their vehicle. Machine learning therefore is a key enhancement that needs further improvements today.

The scenario also mentioned voice notification, the AI capabilities of the CPA, which comprise core technologies like STT, TTS, NLP and machine learning, which allows the system to communicate with the driver. Event management combined with big data allows the CPA to detect the vehicle is moving and then combined with the analytics above predict where the vehicle is going. Of course, outside information like traffic and weather with higher map accuracy are also needed, so the proper ecosystems of merchants and third parties are needed for interfacing with external data sources. A CPA at the highest level therefore uses the capabilities of big data, AI and ecosystems to deliver a personalized mobility experience.

This generation of cognitive vehicle is where all the hype is today, with an automated vehicle without a driver always being in the loop and a CPA that's completely integrated into the personal life. Figure 2.24 shows a futuristic view of a digital cockpit or what we might see in the near future, if we are sitting behind the wheel. Maybe in the case of fully autonomous vehicles, we would prefer to be sitting in a lounge chair or at a table discussing issues with friends.

2.6.1   Solution Architecture

The three major visible changes in the solution architecture of the Generation 3 between today and 2022 are

1. the consolidation of the vehicle capabilities into the digital cockpit,
2. the extension of the service domain navigation to the mobility cloud, and
3. the enhancement from analytics capabilities to AI

Fig. 2.24 A potential digital cockpit of the near future where technology is getting a human touch

(see Fig. 2.25). This generation combines all the machine learning capabilities to enable the autoMOBILE to fully function as an automated vehicle and a personal assistant. This sounds simple to implement, but the complexity is in the integration of all the vehicle data, customer data, and the ecosystem of merchants. This is the ultimate goal for the cognitive vehicle platform for Generation 3 (2022).

2.6.2 Vehicle

In the vehicle layer, we can see right away that the capabilities 'head unit' and 'instrument cluster' are consolidated into the new capability 'digital cockpit.' Beyond that, it is more the services domain 'Navigation' with the new high-definition maps and the cognitive edge capability that are becoming key in taking the cognitive vehicle to the next level. The high-definition maps are primarily managed in the mobility cloud, but updates are beamed into the vehicle and used locally.

Digital Cockpit
Sure the cognitive edge will play an important role, but this could be the digital area Apple and Google are more interested in than actually creating the entire customer experience for the vehicle interior. There is no separation anymore between head

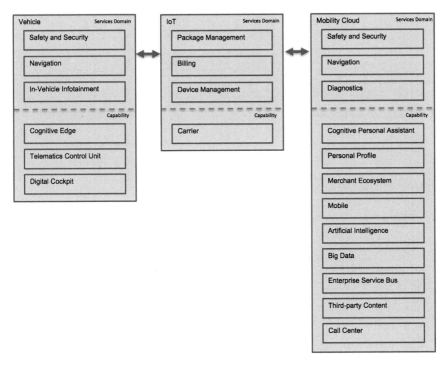

Fig. 2.25 The solution architecture of service domains and capabilities for Generation 3—cognitive vehicle (2022)

unit, head-up display, instrument cluster, etc. It is one operating system integrating the vehicle interior experience. But the digital cockpit is not just more and bigger displays. Instead of notifying the driver of an event visually on a display, the steering wheel could vibrate.

For example, Elon Musk is delivering the Tesla Motors Model 3[22] without an instrument cluster or head unit and is focusing completely on a digital cockpit where the interior should "feel like a spaceship". Similar concepts are happening at BMW with vision Next 100, which is only a concept car.[23] Another example is the Acura Precision Cockpit.[24] These examples show how different the interiors will be in the future and how important events and notifications will be. What objects are around the vehicle, where they are now and where they might be going, events to warn the driver about these things happening all start coming into play.

---

[22]"Tesla—Model 3" http://www.tesla.com/model3. Accessed: July 26, 2017.

[23]"BMW Group—The Next 100 Years—Brand Visions" http://www.bmwgroup.com/en/next100/brandvisions.html. Accessed: July 26, 2017.

[24]"Acura Future Vehicles—Precision Concept" http://www.acura.com/future-vehicles/precision-concept. Accessed: July 26, 2017.

Regarding the merchant ecosystem, events about POIs that are personalized can be displayed in all-new ways, a recent example is where Panasonic showcased an autonomous cabin of the future.[25] These new types of suppliers and ecosystem are difficult for traditional OEMs to deal with. But these new participants in the market that make products that our more focused on customer experience, integrate digital content, and connect ecosystems may be in a better position to deliver these types of vehicles and services.

Example operations regarding the digital cockpit are:

Action	URL	Description
GET	`/device/{id}/content`	Get content from a device
POST	`/device/{id}/display/content`	Display content from a device
PATCH	`/device/{id}/content`	Update content from a device
GET	`/device/{id}/profile`	Get personality profile from a device
POST	`/device/{id}/display/profile`	Adjust vehicle interior environment
PATCH	`/device/{id}/profile`	Update vehicle interior environment
GET	`/vehicle/{id}/road`	Get road surroundings from vehicle
POST	`/vehicle/{id}/display/road`	Display road surroundings
GET	`/device/{id}/area`	Get environment content from cloud
POST	`/device/{id}/display/area`	Display environment content from cloud
GET	`/vehicle/{id}/notify`	Identify best device to notify driver
POST	`/device/{id}/notify`	Send notification event for driver to device
POST	`/device/{id}/identify`	Identify passenger by device
POST	`/vehicle/{id}/identify`	Identify all passengers inside vehicle

2.6.3 Mobility Cloud

Most capabilities previously introduced are staying the same while big data, the merchant ecosystem, and third-party content will increase in volume and frequency of use. But the previous capability 'analytics' will be enhanced to AI to provide the technologies and approaches maturing the cognitive vehicle with the Generation 2022. The high-definition maps are defining the new service domain navigation in the mobility cloud layer.

Navigation

This service domain is completely different from what we have introduced as call centers based navigation services in Generation 1 and disappeared in Generation 2.

---

[25]"Go Driverless with Panasonic's Future Car 2025 at CES 2017" https://channel.panasonic.com/contents/19734/. Accessed: July 26, 2017.

It is a fully digitalized service to manage high-definition maps to achieve self-driving vehicles level 3–5 with up to 5 cm of accuracy. In addition to the usual map of where we are located provided by Google today, the high-definition maps will have the accuracy to show not only what road we are on, but what lane we are in. This will be required for supporting autonomous vehicles in the future. With the previous platform Generation 3 that is feasible today, we already introduced adding shared POIs where information from each vehicle like locations, braking, acceleration, wiper on/off is shared so other vehicles can use the information for safety-type scenarios or for something simple like where rain is actually falling to the ground.

There are not many map providers and even less who have the capabilities to build high-definition maps. Known as global players in the navigation space are, for example, Google, Apple, HERE, TomTom, and in some cases even Garmin. But there are also specialized companies for digital maps and navigation systems like Zenrin in Japan. Usually, their systems are either convenient and user-friendly or provide more accurate information from some perspectives than others.

HERE, the former mapping division of Nokia, achieved higher visibility in the automotive industry because of the acquisition by BMW, Mercedes, and Audi in 2015.[26]

Example operations to update dynamic map content and report errors on map extensions within a navigation service are:

Action	URL	Description
POST	/map/feedback	Submit feedback on map item
GET	/map/feedback/{id}	List status of feedback revision
GET	/map/feedback/search	Search for a feedback item
GET	/map/turnpoint	Draw arrow on top of the map image
GET	/map/maptile	Get a map tile image
GET	/map/basetile	Get image shows map and streets, but no labels
GET	/map/streettile	Get image contains only street lines and labels
GET	/map/venue	List of venues in given geo area
GET	/map/roadsign	Create image for road and traffic signs
GET	/map/version	Get map version details

More comprehensive are for example the HERE REST APIs & Platform Extensions.[27]

[26]"BMW, Mercedes, and Audi are joining forces to take on Google" http://www.businessinsider.com/bmw-mercedes-audi-here-map-data-2016-9. Accessed: July 26, 2017.
[27]"REST APIs from HERE Maps—HERE Developer" http://developer.here.com/develop/rest-apis. Accessed: July 26, 2017.

Merchant Ecosystem

Previously in Generation 3 of today, it was more about core components for managing the ecosystem of merchants. In 2022, the platform will exchange more data across the merchant ecosystems to create customer values as Michael Gorriz said[28]:

> As long as customers are given a choice, see the benefits and are asked for their agreement, they are more likely to share their data.

The CPA has moved from command and control to being connected to ecosystems and providing personalized information to a driver. This allows the driver to ask more types of questions in a natural way. Examples include

- "when will I arrive at my destination?"
- or "how does my adaptive cruise work and when should I use it because it is paid per use?"

Other examples where we analyze driving patterns and make recommendations to the driver

- "would you like to stop at your favorite coffee shop? I can place your order, and you will arrive in 5 min."

With the dialog enhancements in CPAs, the driver could continue the conversation with

- "yes I would like a coffee, but can you add my coupon for a Danish as well".

All of this is also possible with OEMs integrating with merchant ecosystems like General Motors and IBM announced recently.[29]

It is therefore not just about gas or coffee, but the type of gas or coffee the driver prefers. Google also announced similar functions to be integrated into their navigation application, so we will certainly see more of these types of scenarios in the future, and with more competitors.[30]

Other scenarios for electric vehicles are using the data to analyze how we drive and where we are driving to suggest places to stop and to pay for recharging. Eco-driving scenarios also monitor vehicle data and driver characteristics to produce driver scores and even suggest ways we can improve our driving, i.e. hard braking,

---

[28]"Join the data revolution" https://www.digitalnewsasia.com/insights/join-data-revolution. Accessed: July 26, 2017.

[29]"Hello, OnStar—Meet Watson" https://www-03.ibm.com/press/us/en/pressrelease/50838.wss. Accessed: July 26, 2017.

[30]"Google Maps is about to get a lot more ads" https://www.theverge.com/2016/5/24/11761794/google-maps-promoted-pins-announced-ads-coupons. Accessed: July 26, 2017.

acceleration, and overall speed to optimize fuel and battery efficiency. As these types of scenarios improve, privacy and the use of personal information will need to be managed carefully by OEMs and their partners. Customers are willing to provide information when they get a better customer experience in return or have more control and transparency on the information that is being shared. This is still a gap in the existing solutions today, both with regard to the customer's transparency on information that is being shared and collecting more personalized profile-type data to drive an enriched customer experience.

The APIs related to the enhanced merchant ecosystem must identify and aggregate merchants for further interactions related to an event and the collected insights about the targeted consumer. The executed interactions or advertisements depend on the established trust and relationship with the customer. Only some data will be shared with merchants to get individual recommendations and offerings by each merchant on the user-preferred channels for the communication related to the event.

Further example operations for a merchant ecosystem are:

Action	URL	Description
GET	/merchant	List of merchants
POST	/merchant/{id}/data	Share some data with merchant
GET	/merchant/{id}/service	List of services offered by merchant
GET	/merchant/{id}/metric	Data measurements, e.g. earnings, clicks
GET	/merchant/{id}/trend	What are the merchant's top sellers right now
GET	/merchant/{id}/interact	Recommended interaction with merchant
GET	/merchant/interact	Aggregate merchants to interact for an event
GET	/account/{id}/preferences	Get a user's preferences
GET	/account/{id}/channel	List user's channels for ad exchange
GET	/account/{id}/insight	Get insights about potential user's need
GET	/account/{id}/relationship	Get relationship with user

Artificial Intelligence

In the previous service domain 'Merchant ecosystem', we have seen the growing demand to understand how we will get better customer insights to improve relationships. For that we will need to integrate AI into the customer experience and ensure AI technologies are incorporated broadly and in inclusive ways to better interact with everyday users. There are already successes in AI technologies from more accurate speech recognition to better image search to improved conversations, which will further advance by 2022. But AI will go much further and be more useful to all of us if we leverage design thinking (see Sect. 1.4) to open up entirely new AI applications to build the foundation for the next Generation 4 cognitive life.

The AI capability is enhancing and replacing the analytics capability of previous platform generations. The primary added focus is on machine learning, which we introduced in Sect. 1.2.4. For that we need to add APIs related to training and analyzing data models.

Example operations especially regarding machine learning are:

Action	URL	Description
GET	`/analytics/model`	List trained models
POST	`/analytics/model`	Train a new model
GET	`/analytics/model/{id}`	Check training status of model
PATCH	`/analytics/model/{id}`	Add new data to trained model
DELETE	`/analytics/model/{id}`	Delete a trained model
GET	`/analytics/model/{id}/analyze`	Get data the model was trained on
POST	`/analytics/model/{id}/predict`	Request prediction
POST	`/analytics/vision/objects`	Detect objects in images
POST	`/analytics/vision/text`	Detect text within images
POST	`/analytics/video/annotate`	Perform video annotation

2.6.4   How Are These Services Used by Customers?

Let us start with high-definition maps first, we mentioned earlier about the accuracy and detail that will be available in high-definition maps, but how does this directly relate to scenarios. Let us take the example where the CPA in the vehicle says

- "your fastest route to the exit is staying in the middle lane".

Without highly accurate maps that can differentiate the lanes from the road, scenarios like this aren't possible. Next when we start combining this with ADAS scenarios like

- "warning a possible accident ahead I suggest you move into the left lane",

we have both data from vehicle ahead that's being shared, and the detail in a map that suggests to the driver to move into the left lane instead of just slowing down.

In Sect. 1.6 we defined the levels for autonomous driving, most of the scenarios discussed around ADAS are self-driving level 2–4, and for this generation the types of scenarios we will see for level 3 self-driving vehicles. Most of the ADAS scenarios today rely on sensors that can cover an area of about 20 meters around the vehicle. In future, with the cloud, we can start to share the information from other vehicles a half mile or the mile ahead, basically a Waze-type[31] scenario, but

---

[31]"Free Community-based Mapping, Traffic & Navigation App" http://www.waze.com/. Accessed: July 26, 2017.

instead of the incidents being reported by users in their smartphone application, the vehicle sends them in real time. Technically, there are other ways to do this in the industry that are discussed in V2V and V2I scenarios, but we are focusing on doing this through the cloud only.

Another scenario that not only brings ADAS and cloud into play, but data from third parties as well, so when the CPA says

- "the rain is very heavy ahead and water is covering the left lane"

we are now starting to see the combination of the data and the systems required to execute a scenario like this. Microclimate data plus real-time data from vehicles about anti-lock braking or traction control systems can all be combined to create unique customer experiences. More and more, the introduction of third-party content, combined with vehicle data and vehicle controls, lead to more scenarios related to level 5 self-driving vehicles. In the near term, controlled environments like parks and city pedestrian areas can have fully autonomous vehicles to help with traffic and pedestrian flow. Services where we can hop on an autonomous shuttle, ask about favorite places to eat and if the weather will affect the rest of our day are all examples we will start to see in the near future by 2022. It is a different debate about when self-driving level 5 scenarios will be available in general public areas. We believe the technology will be there by 2022, but watching the legal and social effects will play a far bigger role that will take more time. What we will see is that a vehicle will always require a person with driving skills to be inside the vehicle for self-driving operations.

Self-configuring offers new scenarios and new business potential in this platform generation that really has not been seen previously. The first scenario we will look at is related to carsharing, when personal vehicles become carsharing vehicles. Imagine we arrive home from work and we tell our CPA

- "my neighbor wants to use the car, please switch into carsharing mode".

Now the vehicle needs to clear out all our personal settings so our neighbor is not calling all our friends, maybe even turn on a driver behavior application so both are confident no one is abusing the vehicle. Even a simple tracking application can be enabled so in case the son or daughter is using the vehicle we can see where they are and ensure they aren't going out of the area we specified.

Another example for self-configuring is to tell the CPA

- "please download that new horsepower performance update so I can check it out in the mountains this weekend".

This is similar to what Tesla Motors is starting with its autonomous feature, which we purchase and download over the Internet. We introduced merchants and an ecosystem with the Generation 3 of a cognitive vehicle feasible today. With Generation 2022, even the OEMs and maybe third parties will participate

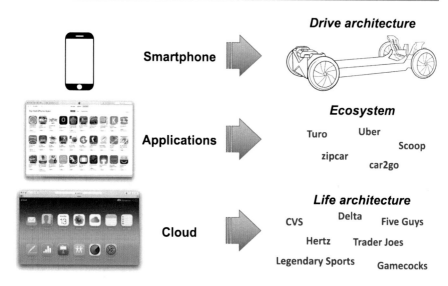

Fig. 2.26 The digitized smartphone context as we know today is defining the reconfiguration of the cognitive vehicle architecture (Photos: Apple)

in ecosystems that bring new services and capabilities to vehicles. Like the app stores that exist for smartphones today, but more geared towards applications that change how the hardware performs, like a smartphone that turns into a bubble level, a compass, a scanner, a personalized credit card, or even a flashlight. In the future, the automotive industry will need these types of software-defined application stores that also change the functional characteristics of the vehicle. However, here we will differentiate between the safety-relevant drive architecture and the life architecture of the vehicle as discussed in [10], (see Fig. 2.26), which illustrates how the digitized smartphone context as we know today is defining the reconfiguration of the cognitive vehicle architecture.

These examples show how software alone can change or reconfigure a vehicle and in future, these changes will be more important we believe than the color or style of the vehicle. As we approach fully autonomous vehicles, the idea of buying a car for handling or looks seems less important than maybe how the interior of the vehicle can be changed to match the driver's needs. We can imagine waking up on a Saturday morning and telling the CPA

- "please set up the interior for four people, golf theme, and download the hole-by-hole aerial shots of the course",

to prepare the autonomous vehicle to pick up the friends for a nice round of golf at the favorite course. Now the experience isn't about the drive and driving, but preparation and accommodation for the friends and the golf experience.

By 2022, technology giants will build on their strengths, partner with others, and enter unique markets where their strengths are amplified. They will not seek to produce millions of vehicles worldwide. In the previously described golf scenario, Apple may be better organized to deliver this type of scenario than a traditional OEM. The interior of the future, the content from the golf course, the networking with friends, and access to personal video libraries seem more like what we are doing with a smartphone than with a vehicle. Apple could partner with a smaller OEM to build the "minimalistic travel capsule" (see [10]). It could use its Apple stores as dealership for sales and configuration, the smartphone or tablet acts as the personal identifier and provides all the personal content, and a new Apple app store could be the new ecosystem to bring in the golf courses and their content.

Last but not least, improving the service experience could be solved through partnerships with existing nationwide service centers or an expansion of Apple stores to service these vehicles. We imagine Apple will certainly fix the worst part of owning a vehicle today and that is the service experience at a dealership. Certainly, as we approach fully autonomous and more self-configuring vehicles, it will be easier to go into a dealership and get a replacement or loaner because through software the user will be able to make that travel capsule just like his own. Or with fully autonomous vehicles, the loaner vehicle shows up at our house and we set the destination for our vehicle to the dealership and off it goes. And for wine lovers, what would be better than to have a fully autonomous vehicle pick us up in the Pearl District and take us down to the Willamette Valley to taste wines in the afternoon. Both cases are high on our list.

Not all OEMs will go in this direction in building such cognitive vehicles and new mobility concepts related to the autoMOBILE, brands and priorities will be different. Regional and global factors will influence these decisions as well. As we think about all the press around Apple, Google, Uber and others getting into the automotive market, first we need to understand what market opportunities will exist in the future and in what regions, and what shares of the market these opportunities will really offer.

For example, the demand for urban mobility will further increase significantly and will continue to be a key factor for the change in the automotive industry. Berylls's study on urban mobility[32] shows that the global mobility budget was €3584 billion in 2015. About 83% are private passenger vehicles, while mobility concepts (including taxi) are below 5%. But in the next 20 years €2531 billion is predicted as the new market for new mobility concepts what will be about 32% of the global mobility budget then.

---

[32]"Berylls's study on urban mobility" http://www.berylls.com/en/informationen/news/170623_mobilitaetsstudie-2035.php. Accessed: July 26, 2017.

Before the entire urban mobility market changes, the first markets will most likely be around college campuses, mixed-use developments, theme parks, and urban developed areas. Many of these markets vary by region. Portland, for example, has the Pearl District and Tokyo has Midtown, both examples of mixed-use developments that could benefit from self-configuring vehicles used by customers who are traveling either locally or outside the cities. The advantages of mixed-use developments, which are areas where people can work, shop, be entertained and live all in a defined area. The trends show that people in these areas are not buying vehicles. However, once a week or once a month they may need a special vehicle for a short time for the weekend. Traditional thinking might result in a rental car, but going back to a personalized experience and self-configuring, these options are more appealing than a rental car. Another reason we like these scenarios is that the developer typically has the money and incentive for these types of services, where cities tend to be more like government projects with less budget. Why did Uber succeed in New York City and Washington DC? For a little higher price, you were provided a more personalized experience, which won out over standard cab service. The same is true for college campuses, which is certainly an incentive to bring students to their school. Students really don't need vehicles that often as well. One area we will not discuss explicitly is 'Smarter Cities.' Sure, some of the concepts and ideas will be part of Smarter Cities, but the general categories covered in this area are much broader, involve local governments and are more complex.

2.6.5 How Far Is It from 'My Cognitive autoMOBILE Life'?

We are assessing the Generation 3 platform (2022) by the foundation of the six self-enabling themes (see Sect. 1.6) and illustrate the score in Fig. 2.27.

If we go back to our framework, let us see how the Generation 3 cognitive vehicle in 2022 measures up. We would score

Self-integrating	4, for real-time content integration across different ecosystems to create a unique customer experiences.
Self-configuring	4, for providing a full personal profile to adapt vehicle and environment for multi-use scenarios.
Self-healing	4, for leveraging the predictive diagnostics to remotely update and repair the vehicle by a virtual domain technician, so a physical visit for maintenance is becoming unnecessary in many cases. Even traditional recalls are turning into remote software updates to reduce warranty costs and improve the overall service experience.
Self-learning	4, as the cognitive edge and CPA matures and more personalized scenarios are achieved by learning more about the drivers and passengers, a path towards real general AI solutions.

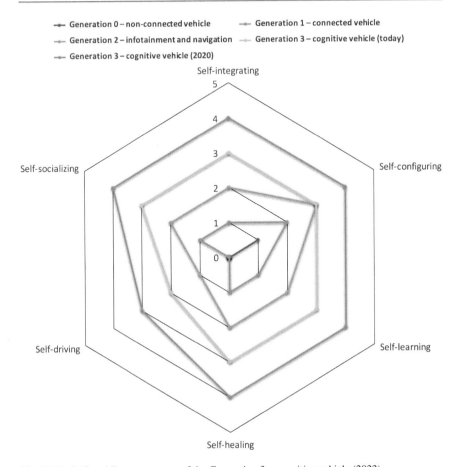

Fig. 2.27 Self-enabling assessment of the Generation 3—cognitive vehicle (2022)

Self-driving   is maturing and moving up to 3 by taking the driver out of the
                    loop for the first time, but with preparation to intervene if nec-
                    essary. The future of fully autonomous vehicles is approaching,
                    but in the short term a lot of unique market opportunities are
                    evolving for both OEMs and new participants.

Self-socializing   4, for ad-hoc socializing of the passengers as well as the vehicle
                    through V2V and V2I as drivers become passengers and can
                    focus on activities other than driving.

We are seeing that even in 2022 we will not achieve any self-enabling theme at the
highest level 5.

2.6.6 How Relevant Is It Across Industries?

One of the interesting problems that will face automotive and other industries as we start to learn more about our customers is how and when to present the information to them. Do we really want Alexa talking every 5 min and saying

▶   "I predicted you need more shampoo, would you like to place an order?",

a simple example of the fine line between a recommendation and an annoyance. This is where the HMI and what we are seeing now with a more advanced and trusted CPA will become critical in product design. How much, when and how do we present information to the user? These are the major challenges that face automotive and other industries with connected products.

We talked before how a CPA is a combination of big data, ecosystems and AI, but is this the correct approach for every product or is the CPA just another part of a multichannel user experience? If we look at Amazon and Echo today, is Echo the product, or just another channel into the core Amazon business of selling online products? Alexa is the CPA behind Echo and as Amazon tries to grow their business around Alexa, will it always be thought of as a channel into Amazon. Will it be trusted to do anything else? Only time will tell and in Chap. 4 we will look at the overall CPA market and see where Amazon stands.

An interesting case to look at, which we will explore further in a cognitive life. However, to explain HMIs and the options available, Amazon is certainly doing a good job in their multichannel architecture to drive consumers to their core business. However, through the web, smartphone applications, and now Alexa, they create a multichannel architecture to drive a more personalized experience no matter what platform we are using. Where is a CPA a good choice as the HMI? We already discussed automotive and overcoming driver distraction issues, but how about other products, and what factors drive the HMI decision? The two main factors are the microphone and the background noise, because in many noisy environments talking to a CPA may be difficult even with a good-quality microphone in the proper location.

Typically, when people talk to a call center, regardless of the industry, this is usually a first implementation example for a CPA or chatbot to enhance customer engagement and even support human-based conversations when they turn sour to find opportunities to improve customer service scripts and dialogs. In reality, call centers can be quite expensive and if we can direct people to an automated system that supports NLP and allows the customer to more naturally talk about their problem, the created solution is better.

Recent examples like Duo[33] where a CPA was put into a mirror to be used in home automation, hotels or convention centers. Usually the hands are busy and cannot grab a smartphone. The mirror is then a quick and easy way to get

---

[33]"An AI computer for your home" http://duo.computer. Accessed: July 26, 2017.

information about a facility, the surrounding area, weather, etc. But the mirror is just another IoT device inside the personalized network. In 2016, IBM and Softbank announced Pepper,[34] a concierge robot for branch offices to answer basic questions and then connect the customer to the best expert to address the questions in detail. In the announcement, they discussed other market areas that can be addressed where kiosks or tablets are typically used today. An interesting trade-off that will require more analysis and that we will explore further in the next Generation 4 for the cognitive life is the decision to connect the personalized products to its own ecosystem or another ecosystem like Alexa to gain the values a customer expects.

An open question that nobody can certainly answer is

► which industry will lead the marketplace of all the app stores?

The leading smartphone technology companies like Google and Apple are already in a very strong position. The electronics industry with companies such as Samsung has a strong focus on enhancing their position. The OEMs are using their strength in the design of the vehicle to establish their own customer-centric application stores. Last but not least, financial services companies are also starting to develop application stores that support the value and flow of money in these ecosystems, because they still command customer trust. This trust is what brings us right to the next generation, which we refer to as cognitive life.

2.7   Generation 4: Cognitive Life

Cognitive life, a larger step beyond a cognitive vehicle, a step beyond a personalized mobility experience in our vehicle, but really a personalized experience as we move through life.

> When I go somewhere, I want that place to understand me. I want that place to have context about me, whether I am at home, in a vehicle, at a hotel, on an airplane, in a retail store, enjoying an event at a stadium, or eating out. I want a personalized and unique experience that I can trust.

Each of these places are different, each of these places are represented by different companies and brands. How can they all be connected in a cognitive life. It is approaching the trusted e-motion, which we defined as the highest level of self-socializing (see Sect. 1.7). Actually, the cognitive life demands all six self-enabling themes to get to the highest level, which is centered on a CPA, (see Fig. 2.28).

---

[34]"IBM Watson to Power SoftBank Robotics Pepper" https://www-03.ibm.com/press/us/en/pressrelease/48763.wss. Accessed: July 26, 2017.

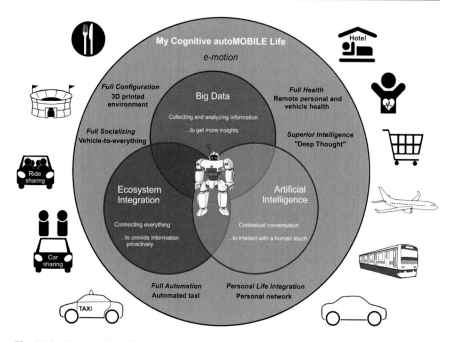

Fig. 2.28 The cognitive life demands all six self-enabling themes to get to the highest level, which is centered on a CPA

Separately, they all care about the same things, customer services, valuable products, and personalized experiences, but how do we possibly put all this together? Can one company do it or will partnerships be required? Digital is likely to be the dominant form of interaction with these customers. Some people may say their smartphone is their cognitive life. It connects them to all their brands and places they are interested in. Some people may say Alexa is defining the cognitive life. Amazon online shopping connects them to all their favorite brands. Right or wrong, let us define some design principles and see how solutions in the market today stack up to being a cognitive life.

To get to some of these principles, design thinking is the approach to start creating this platform Generation 4 for the personalized customer "my way". Good examples are in the book "Designing Your Life" [1]. To start with, the center is one thing that brings us joy or is pleasurable and positive in our life. Let us assume we enjoy being outdoors. Then we started from that as center circle and simply jot down things related to that we enjoy like golfing, running, sailing, etc. (see Fig. 2.29).

Finally, we identify a few words that might be especially interesting to us, and mash them together into, for example, golf trip to the British Virgin Islands. Now we can repeat that exercise for all self-enabling themes to understand 'My Cognitive autoMOBILE Life', like the center says "be more social" or "be more configurable", and see where this practice is directing a company.

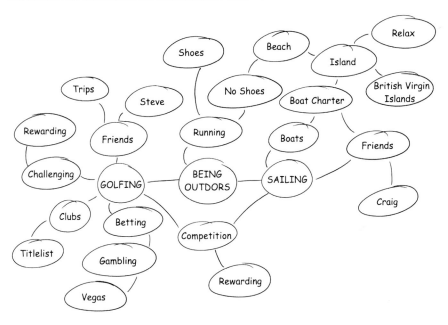

Fig. 2.29 Let us assume we enjoy being outdoors. We then start from that as our center circle and simply jot down things we enjoy related to this like sailing, golfing, running, etc.

Mapping the design thinking outcome on the ecosystem integration for the cognitive life leads us to a set of questions:

1. Who and how many merchants are connected, and are their loyalty programs connected?
2. Next is the digital identity of participating people and how is it shared in the ecosystems?
3. Context, what information and actions are being shared?
4. Value, who gains the potential value in the ecosystem, the consumer should get value, but who else?
5. The architecture, does it primarily support a CPA or is the architecture layer multichannel?
6. Be social, does the ecosystem support social functions, allowing connections with friends and family?
7. Finally, the complex example related to the core business, what is the foundational business behind the ecosystem?

These questions lead us to a set of design principles of how to redesign the platform for the cognitive life. The vehicle is thus no longer the center for the design of the platform as it was up to Generation 3. In this Generation 4, the cognitive vehicle becomes an integral part of the customer's personal network (see Fig. 2.30).

Connected Vehicle   **My Cognitive autoMOBILE life**

The internet is an institution   **The vehicle is an integral part of**
of the connected vehicle   **the customer's personal network**

Fig. 2.30 On the left, the Internet is seen as a basic feature in the connected vehicle as we designed the platform up to Generation 3. On the right, the connected vehicle is an integral part of the customer's personal network defining the Generation 4 platform

Now this set of design principles is defining the personal networks, which makes the cognitive life enjoyable. This is our

health	to be sound in body, mind, or spirit,
retail	where we buy our consumer goods or services,
home	our residence where we sleep, eat, and practice hygiene,
mobility	what directs the flexible motion of people,
city	where we socialize with other people,
vehicle	to transport us or cargo,
smartphone	apps and connecting us to cloud services,
CPA	to perform individual tasks and services.

These eight design principles are thus the foundation to define primarily the Generation 4 platform. Now we conclude with a serious concern. Are the OEMs of today really ready for these types of Generation 4 changes or will new OEMs emerge like Apple and Google? As an orientation for the platform discussion, we picked a couple of non-automotive companies to go deeper into the design principles (see Fig. 2.31). Only time will tell which companies will lead the cognitive life platform, but it will certainly be exciting to watch how the vehicle transforms into "my way", which leads us to the changed focus in the solution architecture.

	Health	Retail	Home	Mobility	City	Vehicle	Smartphone	CPA
Amazon	-	Online	Echo	-	-	-	Kindle	Alexa
Google	Fit	Shopping	Home	Maps	Sidewalk Labs	Interested	Pixel	Assistant
Apple	Health	iTunes	TV	Maps		Interested	iPhone	Siri
Samsung	Health	-	Appliances	-	Smart City	Interested Tier 1	Galaxy	Bixby
Microsoft	Health	-	-	Bing	CityNext	Tier 2	Lumia	Cortana
IBM	-	-	-	-	Smarter Cities	-	-	Watson
Daimler	-	Connections Dealers	-	car2go Moovel	-	Vehicles	-	-
BMW	-	Dealers	-	DriveNow, Scoop	-	Vehicles	-	-
Panasonic	Healthcare	Dealers	Panahome Appliances Products	-	CityNOW	Tier 1	-	-
LG	Google	-	Hub Robot Appliances Products	-	-	Tier 1	LG	Google Assistant
Alibaba	Ali Health	Online	Tmall Genie	Ali Trip	-	AutoNavi Internet Car/SIAC	Yun OS	Alibaba Cloud
Baidu	-	Search	-	-	-	Apollo	-	Duer

Fig. 2.31 The eight design principles define the personal network and are the foundation to define primarily the Generation 4 platform

2.7.1 Solution Architecture

Compared to all previous generations, we now changed the vehicle layer to the "Me" layer to reflect that the vehicle is only an integral part of the customer's personal network and not vice versa. Hence, we will only focus on this new "Me" layer in the solution architecture (see Fig. 2.32) without changing the service domains and capabilities of the other two layers IoT and mobility cloud.

2.7.2 Me

Comparing the 'Me' layer with the vehicle layer of the previous Generation 3 for 2022, we recognize that only the common capability that is left is 'cognitive edge,' everything else is new. We have four new service domains:

1. Cognitive health optimizes the way of healthy living,
2. Online retail as electronic commerce, which allows consumers to buy goods or services directly from a seller over the Internet,
3. Home automation to control and automate lighting, heating, ventilation, air conditioning as well as home appliances such as ovens or refrigerators at home,
4. Cognitive mobility optimizes products and available services with which physical distances can be overcome in accordance with individual needs, and

Fig. 2.32 The solution architecture of service domains and capabilities at the "Me" layer for Generation 4—cognitive life

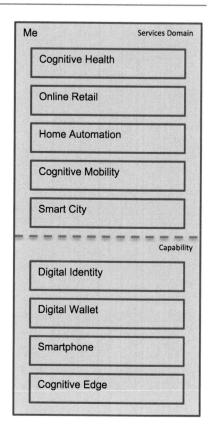

5. Smart city to integrate information and communication technology and IoT technology in a secure fashion to manage a city's assets.

In addition, we define three new capabilities:

1. Digital identity is information uniquely describing an individual, organization, application or electronic device in the cyberspace,
2. Digital wallet refers to an electronic device that allows an individual to make electronic transactions, and
3. Smartphone apps and connecting us to cloud services.

Cognitive Health

For health, we will create two categories. One for wellness/unregulated and the other related medical/regulated. The reason for this is that the FDA has regulations on what is classified as a medical device, which means it requires a different set of qualifications for development, testing and implementation. Most of us are familiar with wellness. Every smartphone has a health application that combined with a wearable can monitor all types of conditions in our body. The CPA is coaching us and keeps a log of how active and healthy we are during the day. Besides measuring a wide variety of medical conditions, e.g. related to anxiety and physical or emotional stress, healthy living is motivating and more fun knowing we are not alone in trying to stay active and healthy. Siri voice calling now: "Craig, you seem stressed. Might I suggest some of your favorite easy-listening music to relax you?"

Medical health application are for people with certain conditions like diabetes that need more specialized monitoring. Other examples include heart rate monitors that integrate with our pacemaker to share information with our doctor. Home health monitoring applications also fall into this category where doctors can remotely monitor patients' vital statistics. We will not document APIs for these types of applications.

Example operations regarding wellness data are:

Action	URL	Description
GET	/account/{id}/stress	Get a user's body stress level
GET	/account/{id}/step	Get user's steps counted
PATCH	/account/{id}/step	Set up individual step challenges
GET	/account/{id}/heartrate	Get a user's heart rate data time series
GET	/account/{id}/food	Get a user's food data time series
GET	/account/{id}/water	Get a user's water data time series
GET	/account/{id}/sleep	Get a user's sleep logs by date
POST	/account/{id}/sleep	Create log entry for a sleep event
DELETE	/account/{id}/sleep/{id}	Delete a user's sleep log entry
GET	/account/{id}/weight	Get a user's body weight logs by date
POST	/account/{id}/weight	Create log entry for a body weight
DELETE	/account/{id}/weight/{id}	Delete a user's body weight log entry
PATCH	/account/{id}/health/goal	Update a user's health goals

More comprehensive APIs related to health data are provided for example by Fitbit.[35]

Online Retail

Let us take as our first example Amazon and Alexa, because at its core is Amazon, an online retailer with cloud computing capabilities that provides the home device Amazon Echo with Alexa as the CPA. It listens to conversations going on at home and when it hears the word "Alexa", it wakes up and then listens for the entire command example "Alexa, order some laundry detergent". Alexa also supports many other commands related to our eight design principles for a cognitive life (see page 146).

---

[35]"Fitbit Web API Docs" https://dev.fitbit.com/docs/. Accessed: July 26, 2017.

Let us select a few examples of merchants: Dominos, Uber, Pandora, WeMo, Ford, Capital One,[36] weather, and sports. If we categorize this ecosystem selection and look at the type of merchants, then we can purchase multimodal travel from companies like Ford and Uber, we cover home automation with WeMo, entertainment with Pandora, dining with Dominos, information like weather and sports, and banking with Capital One. A large ecosystem to cover our daily needs, and Amazon is expanding its software development kit every day. It is also just a matter of time when car sales and trades will be established online, there are already indications that Amazon wants to enter this market.

For personal identification, the user is identified both through the merchants and Amazon itself. Their architecture is multichannel, supporting web, smartphone apps and now Alexa as a CPA. Alexa really does not have any social functions for communicating with friends via chat or email, and it currently does not share the context of the conversation with other merchants. As we look at value, starting with the customer, convenience comes to mind, easy ways to order retail items, no malls or stores, all simply delivered to home, and a multichannel architecture with many ways to do it. Amazon is continuously improving and innovating the logistics as well as the goods delivery. Nowadays, an on-demand manufacturing warehouse or drone delivery, in the future clothing out of a 3D printer or even medicine at home where Alexa turns into the pharmacist.

In the Amazon ecosystem, the merchants gain value through another channel to sell their products. But do they learn more about their customers? Are they maintaining their brand association? We will come back to that in our analysis. Amazon clearly gets the most value, just with Alexa as a new way of browsing through a catalog, they drive more people to their core business of online retail. They learn more about their customers as who connects to more and more types of merchants. From a brand standpoint, Amazon is almost synonymous with shopping and cloud.

Pretty impressive. What is missing outside of airlines, hotels and travel information like traffic? Most of our daily life is covered by Amazon or is it really? When we are in stores, getting gas, at the airport, these places still are not associated with "me", they still may not provide a personalized experience. Other brands, their loyalty programs, their connection to the customer really aren't associated with Amazon. This can actually be a big problem for other brands. If their product is lost in the Amazon ecosystem, then their product becomes more commonplace. Without customer information and brand recognition, it's much harder for a product lines to differentiate.

We cannot talk about Amazon without mentioning China's equivalent, Alibaba. Going back to Fig. 2.31, we can see how closely they match each other from retail to cloud. One major area of difference is in the vehicle with the recent announcement

---

[36]"Capital One is on Amazon Echo. Questions? Just ask Alexa" http://www.capitalone.com/applications/alexa/. Accessed: July 26, 2017.

of entering the automotive market with SAIC.[37] What we really like is the quote from Alibaba's Chairman of their Technology Steering Committee, Dr. Wang Jian: "What we are creating is not Internet in the car, but a car on the Internet. This is a significant milestone in the automobile industry. Smart operating systems become the second engine of cars, while data is the new fuel". What better way to describe the vehicle participating in the ecosystem for a cognitive life? The company's overall strategy is clearly focused on cognitive life, covering all our design principles but one.

The retail ecosystem of Apple with iTunes is different than Amazon. It is the most successful in the entertainment and music industry and brings digital music to the market. Followed by video and other audio functions, iTunes is a content warehouse like no other in the industry. Next to their app store supporting the iPhone and iPad, an army of developers can create all types of applications and functions for their customers.

Retail for Apple is not a strength like for Amazon. Once we leave the digital content space, the Apple store has its limits. Of course, all the major retailers and brands have applications on the Apple store. For financial again, all major banks have apps, but Apple also has Apple Pay for supporting digital-wallet and one-touch payment options, a major convenience for many customers.

The basic APIs for online retail are mostly already defined by the previous platform generations in the service domains 'Package management' and 'Billing,' and the capability 'merchant ecosystem.' Generation 4 promotes a shopping experience to understand buying behaviors, deliver staple foods into the refrigerator automatically based on the personal profile, build loyalty, and inspire a cognitive life.

Action	URL	Description
POST	`/merchant/upload`	Upload model for 3D printing
GET	`/account/{id}/personality`	Get insights about a user personality
GET	`/account/{id}/network`	Get personal network of a user
GET	`/account/{id}/interest`	Get current interests of a user
GET	`/account/{id}/motivation`	Get current motivations of a user
GET	`/account/{id}/emotion`	Get current emotions of a user

Home Automation

Home automation has had both brand recognition and differentiation problems for years. Most likely, a home has mixed appliances from Whirlpool, Electrolux, Panasonic, LG, etc., making it harder to have one home automation ecosystem

---

[37]"Alibaba unveils its first smart car, and it's available for pre-order now" https://techcrunch.com/2016/07/06/alibaba-unveils-its-first-smart-car-and-its-available-for-pre-order-now/. Accessed: July 26, 2017.

and support mixed brands. This is where Amazon and Google with their home automation solutions are stepping in. Major brands can still keep their influence by developing their own applications. These services can be very specific to their appliances like a remote user interface to a tablet or integration of appliances with cooking and recipe sites. Generic home automation solutions will always have only the basic functions for the appliance like on or off, temperature settings, etc., but only a brand can create the unique and personalized experience. More and more schemes are getting standardized to improve generic home automation, like a lighting scheme to change brightness or a thermostat scheme to manage temperatures instead of just the basic on and off functionality.

Let us look at Google Home powered by Google Assistant, Google's version of a CPA. Google of course also has an operating system used on most smartphones, an in-vehicle operating system used in vehicles, and their core business related to Internet search and advertising. Similar in a lot of ways to Amazon and Alexa, but when it comes to home automation, Google with more access to the personalized smartphone, calendar, email etc. can start to handle more personal planning-type functions. Similar on the ecosystem front, where Google interfaces with many other merchants and devices.

Google Home partners like Sony, Philips, WeMo, and many more are able to control most types of appliances, televisions, and devices in many US homes. Google Home is not popular globally yet, but the company may be making announcements and investments in this area as well. We don't see any lodging partners yet, but entertainment and information are clearly Google strengths. This is showcased in the system. With music, television and the ability to answer questions about nutrition, local POIs, weather, personal calendar, fast facts and more. No banking partners yet, but access to financial markets and information is available as well. Personal identification with a Google email or other email account is used across all their platforms. Additional merchant identification will be needed as well. Google does a very good job of delivering context across the platform, but merchants and others don't see the benefit. On the architecture side, Google supports a multichannel experience as well, web, smartphone, home automation, and in some ways vehicles, clearly one of the leaders and visionaries in this field.

Social is not a Google strength. None of the features are built into Google Home yet, but certainly an area they will be investing in as well. The value for Google is in learning about people to drive more advertising revenue and in the long run more targeted advertising revenue. The customer is getting value in the home, like Amazon related to convenience, but Google has high visibility for providing high value-added services in return for the data they receive.

Apple has both Apple TV and Apple HomeKit. Apple TV allows the customer to bring the digital content from any of their devices or iTunes to the TV, and HomeKit has the ability for home automation devices to be connected to a common ecosystem.

Samsung, like Apple and Google, is also in the home automation business. They have yet to announce a device like the Amazon Echo, but with their smartphone integration and soon to be Bixby integration, we just take our smartphone out of our pocket and speak any command about a device in our home that we need. Don't forget they also manufacture appliances, televisions, refrigerators and other home products, so from a device and hardware standpoint, the company is a leader in the industry. Samsung has new offerings for their own products but also the ability to integrate products from other providers as well.

Example operations regarding home automation are:

Action	URL	Description
GET	/house/{id}/layout	Get layout of a house
GET	/house/{id}/device	List devices at a house
POST	/house/{id}/activity	Monitor motion events within user-defined areas
GET	/device/{id}/software	Get software version
PATCH	/device/{id}/software	Update device software
PATCH	/device/{id}/eco	Activate/deactivate device eco mode

More comprehensive APIs related to home automation are provided, for example, by Samsung.[38]

Cognitive Mobility
Google certainly has a focus on personalized movement through life. Having Google Maps and the functions it offers for planning all types of travel, knowing where we are, getting the personalized experience outside the home and more into our life. Google connects to airlines for flight statuses, can get traffic from its map service, is embedded in vehicles as an operating system, and supports Android Auto for in-vehicle entertainment and navigation. Also look at Google Maps, certainly the best on the market when it comes to smartphone navigation, not just in vehicles but in cities for trains, subways, traffic information, and route planning across all means of transport.

Google has made lots of investments into this area, first to offer free navigation on a smartphone, free with StreetView, free with satellite view, all convenient features that differentiate their products from others. Their traffic and route planning service is outstanding in the market as well, the first with multimodal travel planning that includes the train schedules in many global metropolitan areas like Tokyo as well. Many enjoy the service they provide as they are traveling around the world. A real global experience with only some country exceptions and the customers give away whatever data Google needs and wants.

Google's strategy is to get more into the vehicle, which will improve the customer experience. Google was also one of the first in the news, outside the automotive industry, to get interested in autonomous vehicles and other modes of transportation. The maps will play a very important role in the area of autonomous vehicles and

---

[38]"Smart Home Cloud API" http://developer.samsung.com/smart-home. Accessed: July 26, 2017.

Google understands this area well. This may have been enough to push some in the automotive industry to acquire HERE maps from Nokia. It is not just that they make smartphones or that they are the market leader for the smartphone operating system, it's the applications like Google Maps and Android Auto that create the better experience. We have looked at the Google Phone, but with its low volumes it really doesn't have an impact. The Android operating system is the top in the market, but do they really gain a lot of value from that work? We think it is more to keep customers away from the competition than to gain value themselves, other than to drive users to their applications and other services.

Apple positions to bring content, music, and a personalized experience into vehicles may put both these companies in the best position to control the cognitive life. We believe they won't be mass producers of vehicles, but for certain markets and types of transportation the concept of a Google vehicle is driving new thinking.

For travel and multimodal support, Apple Maps isn't up to par with Google Maps regarding content accuracy and richness, but it supports many similar functions. Of course, all the airlines, trains, and ridesharing companies have apps in the Apple store. Samsung follows a different strategy for multimodal travel with the acquisition of Harman to become an automotive tier 1 supplier with deeper integration into the cognitive vehicle and CPA markets. Samsung also supports Android Auto or integration of core smartphone functions into the vehicle. With their announcements related to Bixby, their ecosystem for information related to news, weather, and sports is similar to everyone else's. Where Bixby will look to differentiate is regarding context. Third-party application providers will be able to share context from their applications into the Bixby ecosystem, which can become very powerful and enrich the customer experience. It will be interesting to watch these types of features and functions being developed, a first in the industry.

Example operations regarding cognitive mobility are:

Action	URL	Description
GET	/map/parking	List parking spots for an operation area
GET	/map/vehicle	List available vehicles for an operation area
GET	/account/{id}/current	Get real-time context about a user's trip
POST	/travel/request	Request travel
GET	/travel/{id}/price	Price range estimate
GET	/travel/{id}/time	Estimate when travel can begin
POST	/travel/{id}/booking	Booking travel
GET	/vehicle/{id}/rental	Get rental terms
POST	/vehicle/{id}/booking	Book a vehicle for a user
DELETE	/vehicle/{id}/booking	Cancel an existing booking
POST	/devliver/request	Request a delivery

More comprehensive APIs related to transport experiences are provided, for example, by Uber.[39]

Smart City

> A *smart city* is one that looks to use technology to make an urban area more affordable to lead meaningful and fulfilled lives.

A smart city harnesses ecosystems and data by connecting and optimizing for example:

- public transit and physical infrastructure,
- energy efficiency,
- safety and security,
- environmental factors,
- sustainable buildings,
- tourism, and
- healthy economic, social, cultural development.

The smart city is an example of building towards a cognitive life due to it being an entire ecosystem on its own. Some of the difficulties smart cities have with standards and interoperability will apply to cognitive life as well. But the advantages for tourist and the people living in large cities are only gained through these factors. If the bus schedule is not available or up-to-date with the train schedule, a seamless travel experience is much more difficult. The mix of private and public systems also presents the same difficulties, especially in the new mobility ecosystem. Ridesharing, carsharing and parking companies need to integrate with public systems to create seamless travel experiences, and with integration into vehicles the exchange of information can be even more valuable. We mentioned before that Google Maps is a leading example of how all this information can come together and create a personalized experience. No matter what factors contribute to a smart city, APIs create the interoperability to connect them all to create valuable functions by leveraging the massive amounts of urban data being generated and shared across the city ecosystems.

Example operations regarding smart city are:

---

[39]"Uber Developers" http://developer.uber.com. Accessed: July 26, 2017.

Action	URL	Description
GET	`/city/{id}/event`	List a city's events
GET	`/city/{id}/event/{id}`	Get event details
GET	`/city/{id}/place/{id}`	Get place details
POST	`/city/{id}/issue`	Report a civic issue
GET	`/city/{id}/case`	List of matter being decided within orgs
GET	`/city/{id}/action`	List of actions taken by decision-makers
GET	`/city/{id}/person`	List of real persons, alive or dead
GET	`/city/{id}/organization`	List of groups with a common purpose
GET	`/city/{id}/membership`	List relations between member and orgs
GET	`/city/{id}/post/{id}`	Get people holding the post in an org

More comprehensive APIs are provided by the City Service Development Kit (CitySDK), co-funded by the European Union, which is an initiative to increase smart city development.[40]

Digital Identity

> *Digital Identity* is information uniquely describing an individual, organization, application or electronic device in the cyberspace.

The digital identity of a person can be uniquely characterized by combining many data points

Expression: What I say
Publication: What I share
Profession: Where I work
Opinion: What I like
Location: How and where to join me.
Reputation: What is being said on me
Passion: What I favor
Certificate: Who can certify my identity
Purchase: What I buy
Knowledge: What I know
Avatar: What represents me
Contact: Who I know and who knows me
Biometrics: What characterizes my body

---

[40]"CitySDK—City Service Development Kit" http://www.citysdk.eu. Accessed: July 26, 2017.

It is all about the accuracy of and confidence in the data. India has the world's largest biometric ID system in Aadhaar. A person does not need cash or credit cards to shop for basic needs. Just place the thumb on a fingerprint scanner and a payment straight from the bank account is authorized for the shopping. Such a use case leads to the question whether a government or a company like the shop should own and manage the system with the digital identities. In reality, both will exist, some optional and some mandatory.

An Apple ID is used as personal digital identity within their ecosystem, but this is not shared with external providers. Similar to context, smartphone applications are not allowed to share context. This may be something to watch with Samsung's recent announcements,[41] and how sharing context improves the user experience with Bixby.

From the home point of view, we are moving through life in and around our home. Amazon has us identified and covered, but outside of that not so much. This takes some type of identification, some type of location-based system so the merchants or places can understand we are with Amazon. Will their announcement with Ford[42] of putting Alexa in the vehicle, not embedded but through their mobile application, start to open up their presence outside the home? Will Amazon expand more into these other areas?

Only time will tell how the ownership of digital identities will further evolve and what combination of data points will be necessary to uniquely identify a person for different levels of security.

Well known are the OpenID Connect APIs what we will not list up here.[43]

Digital Wallet

A *digital wallet* is a cash management system that allows an individual to conduct electronic transactions using an electronic device. Most common is the use of mobile numbers instead of account numbers.

The smartphone impacts traditional banking because users can shop, pay bills, send and receive payments using a wallet balance without a bank account. But this does not mean that banking will disappear. Even this industry is massively disrupted by digitization. There will always be a need for keeping our money safe, growing it, investing it, borrowing it, or moving it. Only how we are doing it is changing.

---

[41]"Bixby : A New Way to Interact with Your Phone" https://news.samsung.com/global/bixby-a-new-way-to-interact-with-your-phone. Accessed: July 26, 2017.

[42]"Alexa in the Car: Ford, Amazon to Provide Access to Shop, Search and Control Smart Home Features on the Road" https://media.ford.com/content/fordmedia/fna/us/en/news/2017/01/04/alexa-car-ford-amazon-shop-search-home.html. Accessed: July 26, 2017.

[43]"OpenID Connect" http://openid.net/connect/. Accessed: July 26, 2017.

Apple Pay is a mobile payment and a global digital wallet service for Point of Sale (POS) online and for in-app payments. It is compatible with Apple devices and near field communication (NFC) POS terminals. Google enables mobile wallets on all Android devices across OEMs with KiKat, which is also compatible with NFC POS. Samsung uses magnetic secure transmission, which turns existing magnetic readers into mobile contactless receivers enabling wider merchant acceptance than Apple Pay. It is compatible with their devices only.

Example operations regarding digital wallet are:

Action	URL	Description
GET	`/account/{id}/wallet/{id}`	Check available balance
POST	`/payment`	Make a payment
POST	`/payment/{id}`	Payment details, e.g. approved/declined
POST	`/payment/{id}/void`	Void a payment transaction
POST	`/payment/{id}/refund`	Refund a processed payment

More comprehensive APIs related to digital wallet especially with merchant integration are provided, for example, by Mastercard.[44]

Smartphone

The smartphone is at the core of our entire analysis. It will be open how these multipurpose devices will be further evolving and if they are really better than products that perform only one function. But for a cognitive life what better way to connect to an application and brand ecosystem. A few comments on the future features of smartphones and the trade-offs with single-function devices. Take, for example, NFC for payments. Today, single-function devices like the Exxon Speed pass already do this. But for convenience, how many additional devices do we want to carry. For the smartphone display, flexible options are being investigated, but it's not sure yet when these will be available or what the advantage beside not breaking so often will be. Augmented reality is a nice feature, but will it really change the overall interaction and HMI for the smartphone core functions? Probably not. Voice HMI is the best improvement, but it will not completely replace touch due to speed and responsiveness. Multipurpose devices like smartphone will therefore be around for a while. We see some core capabilities that a smartphone provides that help enable applications in this ecosystem. Starting with cellular, Wi-Fi, Bluetooth and NFC, all are core wireless technologies that enable multiple types of communications and integrations for applications. In some cases, the CPA is available for application integration, but this will probably change more in the future. If an application wants to learn beyond voice recognition, the developers may need to look at other technologies as well. Of course, the camera and microphone

---

[44]"Mastercard Developers—Masterpass" https://developer.mastercard.com/product/masterpass. Accessed: July 26, 2017.

are used more frequently in applications. NFC is also used in the area of payments and supporting a digital wallet. We are starting to see in some cases where a second channel besides the smartphone might be necessary for an additional level of security, like smart keys for vehicles. Fingerprint recognition is the main security feature on smartphones as well, used for access control and authentication. It is great for easy access to applications instead of typing a password. Last but not least, all the social and interactive functions like the cell phone and video messaging.

Example operations regarding a smartphone are:

Action	URL	Description
POST	`/device/{id}/display`	Turn on display
POST	`/device/{id}/camera`	Take a picture
POST	`/device/{id}/microphone`	Turn on audio streaming
POST	`/device/{id}/speaker`	Output audio through speaker
POST	`/device/{id}/nfc`	Connect NFC
DELETE	`/device/{id}/nfc`	Disconnect NFC
POST	`/device/{id}/nfc/{id}`	Share NFC ID
POST	`/device/{id}/bluetooth`	Connect Bluetooth
DELETE	`/device/{id}/bluetooth`	Disconnect Bluetooth
POST	`/device/{id}/wifi`	Connect Wi-Fi
DELETE	`/device/{id}/wifi`	Disconnect Wi-Fi
POST	`/device/{id}/cellphone`	Connect cell phone
DELETE	`/device/{id}/cellphone`	Disconnect cell phone
POST	`/device/{id}/video`	Video call
DELETE	`/device/{id}/video`	Video call
POST	`/device/{id}/text`	Send a text message
GET	`/device/{id}/text`	Retrieve a text message

2.7.3 How Far Is It from 'My Cognitive autoMOBILE Life'?

We are assessing the Generation 4 platform against the foundations of the cognitive vehicle defined by the maturity of the self-enabling, (see Sect. 1.6). We score, (see Fig. 2.33):

Self-integrating	5, content is only increasing and becoming more real-time, more personal and creating unique and better customer experiences.
Self-configuring	moves up to a 4, the main focus of level 4, but fundamentally using software to configure and personalize vehicles for multi-use scenarios.
Self-healing	is moving up to a 5, even though we didn't mention a specific new scenario about self-healing, OEMs are constantly

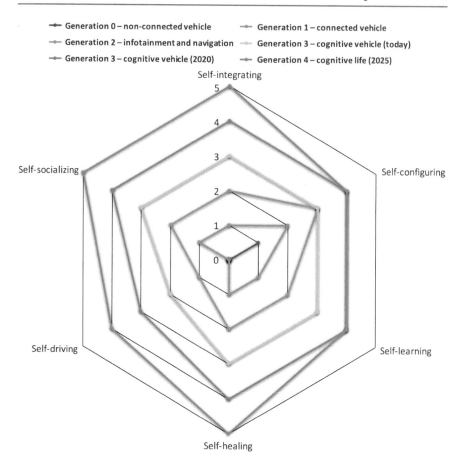

Fig. 2.33 Self-enabling assessment of the Generation 4—cognitive life

	improving this area mainly to reduce warranty costs and improve the overall service experience before and after customers are at the dealer.
Self-learning	is moving up to a 4 as the cognitive edge matures and more personalized scenarios are achieved by learning more about the drivers and passengers.
Self-driving	is also maturing and moving up to a 3 with the future of fully autonomous vehicles approaching, but in the short term many unique market opportunities for both OEMs and new participants.
Self-socializing	moves to a 5 as drivers become passengers and focus more on activities than driving.

2.7.4 How Relevant Is It Across Industries?

This section on cognitive life is probably the most relevant to other industries. Either yours is a company creating the systems to support me at the center, or yours is a company that wants to participant in the ecosystem with differentiated services. We have talked enough about Amazon, Google and Samsung so what about everyone else. How can the same technologies drive us to differentiate our products, make them cognitive or adding a CPA to create a personalized experience? First, let's talk about companies that can help or enable us to do that and then go through a few examples.

Microsoft has the capabilities in these areas as more of an enabler of cognitive solutions and CPAs with Cortana and the cloud. One point about Cortana. It is a CPA, which is available for download on most smartphone app stores, on Windows 10 personal computers and laptops, but their market share as a CPA is small. Microsoft does want customers to use their CPA to enable their products with voice and cognitive technologies. IBM also has similar capabilities with Watson and the cloud to enable customers to develop cognitive solution.

Both IBM and Microsoft have competing products in the cloud category from mobile to personal profiles. IBM's recent announcement with GM related to a merchant ecosystem is one area of differentiation. For the IoT, Microsoft has dynamics for package management and billing, with both IBM and Microsoft having the capabilities regarding device management. On the vehicle side, Microsoft has an automotive operating system for a while used by Ford Sync and others, but recently they haven't been as much an industry player. Microsoft has a cloud strategy to enable customers, so does IBM. One additional capability not previously discussed regarding the *cognitive vehicle* is productivity, and Microsoft's ability to integrate with their office products for scheduling and contacts creates a better customer experience and more related to a cognitive life. So what we will typically see with these companies are custom-built CPAs, i.e. Watson for Oncology[45] or cognitive rooms with Harman,[46] or Microsoft[47] with BMW and Nissan. Because in comparison to a standard off-the-shelf CPA, certain situations and markets require very specialized data that are outside the scope of Yelp, news, weather and sports.

Now let us look at location awareness in general and how important it is in a cognitive life. We talked about some of the scenarios in automotive already, the example of our fuel light coming on and the CPA suggesting our preferred gas station. A simple example, but it shows the importance of location awareness.

---

[45]"Watson for Oncology" https://www.ibm.com/watson/health/oncology-and-genomics/oncology/. Accessed: July 26, 2017.

[46]"HARMAN and IBM Watson Internet of Things Introduce Cognitive Rooms that Bring Connected Experiences to the Consumer" http://www.prnewswire.com/news-relea ses/harman-and-ibm-watson-Internet-of-things-introduce-cognitive-rooms-that-bring-connected-e xperiences-to-the-consumer-300441558.html. Accessed: July 26, 2017.

[47]"Nissan and BMW bring Microsoft's Cortana assistant to cars" https://www.theverge.com/ces/ 2017/1/5/14184140/microsoft-cortana-nissan-bmw-car-integration. Accessed: July 26, 2017.

Another example in location awareness where automotive and home automation come together is the simple example of automatically turning on our lights or the climate control system as our vehicle approaches our home.

Any time we are traveling, location awareness is important as well. What can we eat around where we are? What can we see around where we are? How can we get from point A to point B? There are endless examples. Now what happens when we get somewhere, or we are on our way there, do merchants get that information? Let us take hotels for example, do they know when we are on the way, do they know when we enter the door? Doesn't seem to be a top priority. Hilton[48] recently set up for digital check-in, so we can select our room, check in and then use our smartphone as the room key. Maybe we can enter an estimated time of arrival, but the features are more related to improving the check-in experience—and reducing human resources—than knowing exactly where we are. The key can easily be used for other services in the hotel, to confirm purchases or access the gym. We always bothered for years carrying around the room key to get services in a hotel or how many of those have we collected and thrown out at home?

Seems airlines like Singapore Air[49] are looking more at the check-in process as well than being concerned exactly where we are. When we check in, we can even view and select our favorite movies, so when we arrive at our seat, these preferences are already set up. We really liked this feature compared to browsing 100 or so movies on a crude touch screen to determine what we want to watch.

2.8   Wrapping Up the Architecture of E-motion

Let us start with our foundation for a cognitive vehicle and look back through all the generations and see what we have "learned". First, let us put the criteria into two groups:

1. one with self-learning, self-driving and self-configuring
2. the other with self-healing, self-integrating and self-socializing

Now for the first group, it is interesting that the industry started off strong in self-configuring, probably because they wanted to physically adopt the cars to people, but slowed down once it started to get too personal. Overall, this group didn't reach level 5. If we look specifically at self-driving, which the other two clearly support, even beyond 2025 not all the level 5 autonomous issues will be worked out from regulations to insurance and even under what weather conditions level 5 can work. Autonomous driving will still require a lot of self-learning to help understand

[48]"Let Yourself in with Digital Key" http://hiltonhonors3.hilton.com/en/hhonors-mobile-app/digital-key.html. Accessed: July 26, 2017.

[49]"The SingaporeAir App" https://www.singaporeair.com/en_UK/us/mobile-app/. Accessed: July 26, 2017.

the road, conditions and environment around the vehicle. An autonomous car in the future will even use self-configuring to change its characteristics for its daily commute. For the second group, they have all reached level 5, mainly driven by the digital process that the vehicle will become more like a smartphone and being connected to an extent that it will bring with it the software improvement around healing, socializing and integrating that we enjoy with our smartphones today.

We made our assessments against the foundations for the cognitive vehicle based on the industry as an average, but the reality is each OEM is different. Some OEMs will choice the focus to be on the driver and grow faster in the areas of self-integrating and self-socializing. Other OEMs will choice to focus on self-driving, while others on driving itself and as a result stay more with the physical AUTOmobile. No one way is right or wrong. Each company's strategy will focus on different areas and now they can use our foundational self-enabling themes as a more detailed guide of their journey. The e-motion in our case is mostly about self-socializing, but also influenced by self-learning and self-integrating, and building towards that cognitive life. That doesn't mean that the e-motion is lost from driving, certainly not for BMW with the slogan "The Ultimate Driving Machine". In all cases, data will be the new "driver", with data from all the sensors feeding the AI for autonomous driving, feeding the AI for the cognitive edge and feeding the AI in the mobility cloud layer. The autoMOBILE will truly be data-driven and if customers choose, it can also be a part of their cognitive life.

Coming back to the beginning of the introduction (see page 2) when we asked the question:

►   Do we LOVE our cars?

Section 1.7 showed us that love is trust and happiness. After the exploration of the different platform generations and the evolving role of the CPA, we need to change the question to:

►   Do we LOVE our car-centric ecosystem or do we LOVE our life-centric ecosystem?

That question leads to the major difference between the autoMOBILE and the cognitive life.

In our analysis, Apple more than Google or Amazon is leading in this drive to a cognitive life. Remember our definition

> a step beyond a personalized mobility experience in our vehicle, but really a personalized experience as we move through life,

all that's really missing is context and the personalized experience in the vehicle.

We believe that Google not having strong devices will have a much harder time to control the user experience as they move towards a cognitive life. Similar with Amazon, both Google and Amazon have the devices in the home, but not outside the home, not as they move through life. These are Apple strengths. But let us also look at one of Apple's weaknesses, which is related to context. Apple may change in the future, but context and sharing context across smartphone applications may be something strategically they are looking at. Context is a core factor in user experience on a smartphone, just like smartphone applications can use GPS and share the context of location, it doesn't go further than that.

For example, if we are in the Delta Air Lines application, unless Delta integrates weather, we cannot quickly and easily get the weather at our target destination. With context, Siri for example, could get the destination from the Delta application so when we ask about the weather, it knows the destination we are talking about.

As we mentioned before, Samsung has announced Bixby, which can use context across applications on the new Samsung smartphones. This will be an interesting feature and strategy to watch. The previous Delta Air Lines example may be simple, but context is an important factor in user experience.

Now let's get to autonomous vehicles and the market opportunity for Apple. First, we don't see mass production to the levels of most automotive OEM today, but a focus on markets and opportunities where the customer experience is more a factor than how we get around. College or work campuses are the examples where people don't always need vehicles. Once they are in the vehicle, it is more about content like music, videos, and pictures as they converse with their friends. These types of vehicles for carsharing and ridesharing examples are better suited for customer experience than the everyday commute to work.

Long term past 2030, this may be different, but in the short term, what a great market to attack. Apple currently has CarPlay, but most of the experience is owned by the automotive OEM. Currently, the OEMs are only allowing a few applications, specifically related to entertainment, texting, mapping, and communications for making a phone call. Apple, we believe as the industry moves towards autonomous, will want to own more and more of that experience, beyond just entertainment. It wants to own the personalized experience as well, just like we have with our smartphone today. Combining this with their ecosystem of application providers is a very strong case for a cognitive life in the making.

2.8.1 Different Architectures

On the architecture side, Apple has both a multichannel and multi-device strategy creating a seamless experience across devices with the iCloud. Siri as their CPA is also supported across devices from Apple TV to the Apple Watch. Their core business is devices and hardware, but backed with services and a unique customer experience. All of this drives the unique value that Apple brings to its products and ecosystems, where merchants and third parties gain routes to market but are clearly hidden behind the Apple brand. Customer value is at the forefront of Apple's strategy. They pioneered customer experience, putting the customer over technology, values other companies are only starting to replicate today. Looking at the model, it will be interesting to see whether the customer experience and ecosystem will command the same premium on the price of the vehicle when we get to autonomous vehicles? We will look at this more in our analysis, but certainly a business model and strategy that can be used for other products.

With Samsung and Apple so close in capabilities and strategy, our analysis of Samsung will focus on their broader strategy for cognitive products. Samsung has a unique position in the market because they cover such a broad range of products related to a cognitive life, televisions, home appliances, personal computers, smartphones, and home theaters. We also already mentioned recent acquisitions of Harman and ViV to broaden their portfolio into vehicle components, office products, and advanced CPAs. If we don't know the history, the developers of ViV previously developed Siri, which was acquired by Apple. Samsung is slowly putting together a set of cognitive products that will be in the hands of their consumers throughout the day. One piece that seems to be missing is an extension into the ecosystem of other brands and merchants required to complete a cognitive life. Sure, the smartphone and the applications have a connection to these brands, but will that be enough? Will the introduction of sharing context with Bixby be enough to create an ecosystem that can it create a personalized mobility experience. We talked about these examples before, airlines, hotels, dining, etc. Can we connect and personalize these experiences through the smartphone? Since some of these announcements are just happening, it is hard to say, but strategy tells me this can be a good approach with Harman for their connection to vehicles and the IoT for their ecosystem. The future is strong.

2.8.2 The Technical Challenges of the Cognitive Personal Assistant

First, let us take some issues that came up as we looked at what companies are doing today and others we haven't mentioned. The microphone, a simple yet complicated piece of technology, if not handled correctly will make a CPA worthless. This can potentially be the biggest problem with smartphone providers driving the cognitive life and acceptance of the CPA being on the smartphone. If we cannot talk into the phone in crowded and noisy places, it is basically worthless. It is not just general noise reduction, it's more like when our ears are focusing on only what one person is speaking while other people are talking as well. A Bluetooth headset may be the answer or an extension to the wireless Apple earbuds. Sony has the Xperia Ear[50] in Japan and technically this may help to eliminate the general noise problem. In addition, we might need a headset, glasses or other sensors that the CPA can use to understand which voices we are focusing on and who wants to be wearing all that if it's not part of our regular clothes.

---

[50]"Xperia Ear—Your personal assistant" http://www.sonymobile.com/global-en/products/smart-products/xperia-ear/. Accessed: July 26, 2017.

Maybe the microphone can be upgraded on the smartphone as Amazon has done with the microphone on the Echo.[51] Automotive has taken years to solve the noise problem in the vehicles, but now it's a much better experience to use voice recognition.

Next is a CPA really required, or can the simple HMI aspects of a smartphone, like swipe and tap, etc. make it user friendly enough? In some domains, the CPA might be already starting to surpass the mobile web and in some cases, the experience may even be approaching the quality of mobile native apps. For example, when booking airline tickets and hotels through Hipmunk's bot, we also receive instant advice and recommendations on travel questions.[52]

Well, we certainly think a CPA is required, but from a customer experience standpoint, the integration with a smartphone will be an important design factor. Next is micro location, not location-based services like GPS. Sure, the CPA can identify where we are, pretty much anywhere outside, but once we go into a building or store, it's much harder to do. Apple tried a technology a few years back in retail call iBeacon, which is basically GPS, but inside a building. Not very successful for these and other reasons, along with the cost to implement the hardware at the specific locations we wanted.[53]

As we look at all the CPAs in the market today, a strong ecosystem with lots of data supporting the CPA is required to leverage intelligently. We see the ecosystem behind Alexa with the Amazon shop, we see the ecosystem behind Apple, Samsung and Google with their app stores, and basically all of them are in the home automation market as well. That leaves automotive and with rumors and announced every day, even recently with Samsung testing autonomous vehicles.[54] they all see the value of controlling the customer experience in this field as well.

So do automotive OEMs stand a chance, can they compete with the likes of Apple, Samsung and Google ecosystems empowering the CPA? One of the factors will be brand, and how OEMs can control a brand experience within or around a vehicle. The other, as we have already seen with IBM and Microsoft, the domain of information is specific enough that an off-the-shelf CPA won't compete. Sure, they will still need to partner for weather, news and traffic, but for information related to the vehicle and its health, it's not personal, it's very vehicle-specific domain.

---

[51]"How Amazon Echo Can Hear What You're Saying From Across the Room" http://abcnews.go.com/Technology/amazon-echo-hear-room/story?ID=26740479. Accessed: July 26, 2017.

[52]"Hello Hipmunk" http://www.hipmunk.com/hello. Accessed: July 26, 2017.

[53]"Apple's Failed Retail Revolution: Beacons Still Won't Be Big on Black Friday" https://www.bloomberg.com/news/articles/2015-11-24/apple-s-failed-retail-revolution-beacons-still-won-t-be-big-on-black-friday. Accessed: July 26, 2017.

[54]"Samsung steps up its driverless car challenge to Google, Uber, Apple" http://www.cnbc.com/2017/05/02/samsung-driverless-car-test-approval.html. Accessed: July 26, 2017.

2.8.3   Outlook

In conclusion, maybe the cognitive life will be led by a brand we haven't even heard of today. Remember Apple wasn't even a leader in the smartphone field when they launched the iPhone, but either way, we can see some strong leaders today that we feel are in a pretty strong position. Panasonic, Sony and other electronics companies are great enablers in this market, but do they have the customer focus that's required in a cognitive life? Sure, they will have great cognitive products, but stitching them together into a cognitive life under their brand could prove to be a challenge. Actually, most brands need to be careful not to be commonplace in this new world of a cognitive life, as Amazon and other brands are becoming so strong.

Automotive is a unique market with a focus on safety and the driving experience today. OEMs in the short term will have good reason to have their own CPAs in the vehicle, but watch out for Apple, Samsung and Google. They are all going after this market as it transforms in the next 5–10 years. It is wrong to expect that the AUTOmobile or even the autoMOBILE will define the new ecosystems of the cognitive life where 'Me' is at the center as illustrated in Fig. 2.30.

Multimodal travel and domains closely related to travel like parking are areas of interest for OEMs, but outside of that the OEMs will struggle. Amazon, which is strong in the home and with their approach to an ecosystem, they don't have the devices to compete with Apple, Samsung, Google and Alibaba outside the home. And of the remaining four, Samsung is leading in shared context that drives a more seamless experience across applications. Maybe applications and smartphones won't be what drives cognitive life, but we will be hard pressed to see something else.

References

1. Burnett B, Evans D (2016) Designing your life: how to build a well-lived, joyful life. Knopf, New York, NY
2. Chen F (2006) Designing human interface in speech technology. Springer Science+Business Media, New York, NY
3. Conway ME (1968) How do committees invent? Datamation 14 (5):28–31
4. Frost & Sullivan (2016) Executive analysis of self-learning artificial intelligence in cars, forecast to 2025. Strategic Insight K053-18, Mountain View, CA http://www.frost.com/sublib/display-report.do?id=K053-01-00-00-00. Accessed 26 July 2017
5. Huang C-M, Chang Y-C (eds) (2010) Telematics communication technologies and vehicular networks: wireless architectures and applications. IGI Global, Hershey, PA
6. NGTP Group (ed) (2010) NGTP in a nutshell, Version 1.0 http://ngtp.org/wp-content/uploads/2013/12/NGTP20_nutshell.pdf. Accessed 26 July 2017
7. Perun S, Wedeniwski S (2015) Dynamically linking information in a network. Patent US20170147687, USA
8. Schäffer F (2012) OBD on-board-diagnose in der Praxis. Franzis, Haar bei München
9. Scheer A-W (1990) CIM computer integrated manufacturing: der computergesteuerte industriebetrieb, vol 4. Springer, Berlin
10. Wedeniwski S (2015) The mobility revolution in the automotive industry: how not to miss the digital turnpike. Springer, Heidelberg

Value Chain and Business Models

3

The automotive industry is all about the value chain and business model to build the AUTOmobile product or to disrupt them with the product-service system autoMOBILE (see Fig. 1.3). In the previous Chap. 2 we evolved five platform generations defined by services domains and capabilities towards the cognitive vehicle and cognitive life, which we will now bring into a business model perspective. Our basic principle building the models throughout these generations is, to quote Buckminster Fuller [6]:

> You never change things by fighting the existing reality.
> To change something, build a new model that makes the existing model obsolete.

In that sense, the value drivers and business models have been disrupting the automotive industry since General Motors first released OnStar in 1996. Subscription models were the craze and TSPs were born since most OEMs did not want to invest and build the costly infrastructure to support these types of services. Nowadays that technology has changed dramatically and the cost for implementation has dropped significantly. OEMs are now looking at where to invest and where to get the value, as the new generation of cognitive vehicles is about directly connecting with the customer and providing a personalized experience. The value will be looked at from the customer's point of view, the OEMs' point of view, and the service providers' point of view. Related to the value from the OEMs' point of view, we will introduce a framework for how the internal value of connectivity could be measured. This framework will

1. look at the use case,
2. identify the organization that gains the value,
3. identify who made the investment, and
4. how the *Return on Investment* (ROI) was measured.

© Springer-Verlag GmbH Germany 2017
S. Wedeniwski, S. Perun, *My Cognitive autoMOBILE Life*,
https://doi.org/10.1007/978-3-662-54677-2_3

When we look at a standard organization for an OEM, we will focus on a few of the key organizations related to the development of products and services. The key organizations within the OEM that we will use for the framework are:

1. Product Engineering includes the product teams along with research and development. Each product is typically managed by one team, while research and development spans across the individual product teams providing insights into new and innovative technologies.
2. The Marketing and Sales organization is responsible for the marketing of a product and conducting the research and analysis into what customers want and into their products' contents. The sales side drives the relationship with the dealers and tracks sales targets and objectives for each product.
3. The Quality and Warranty organization is responsible for product quality and monitoring quality throughout the manufacturing process. Warranty management handles defects in parts that can be reimbursed by the manufacture to the consumer.
4. The IT organization is responsible for computer systems, networks, and applications that support the business processes of a company.
5. The After Sales and Service organization is responsible for the product life cycle after the sale to the customer. In automotive, they work with the dealerships' service and repair departments to ensure products are maintained and repaired to the correct standards.
6. The Connected Services organization resulted from mature companies dealing with connected solutions in the market and typically combined cross-functional teams into a single organization responsible for the digital services launched around their products. One advantage to bringing together the different organizations was to provide a more streamlined management process and measurement system for digital services.

Most companies across all industries struggle with internal value sometimes called soft value in consulting when they are putting together a business case for these new connected solutions. Most companies want to see an increase in revenue or a decrease in costs to drive their business case, and will put *key performance indicators* (KPI) in place to measure that, but the difficulty to measure soft value creates its own problem. KPIs for soft values are much harder to define and measure—this is what we believe to be the main driver for the lack of investment in connected solutions. As any OEM in any industry looks at a product-to-digital-services transformation, sometimes the focus needs to be on the soft value, and not just on increasing revenue. The CEOs that were early pioneers and made the decisions and commitments are in a much better position with their customers. These companies are more prepared as products change to adapt to a personalized experience.

One way we can look at OEMs and customer value is going back to our fundamental self-enabling themes for the cognitive vehicle and adding a value factor for each theme. For example, we give a low, medium, and high grade for each theme, assign those corresponding factors one to three respectively, then multiply them to get a value score for the customer and OEM KPI by theme. Since our generations are generic, each OEM can do the exercise to their own specifics, giving their own scores to each category. It will be best to do the exercise for each model or brand of vehicle, if the customer persona we are selling to is different, then this will be required. Specifically as we look at new millennials who may be paying for carsharing rather than buying a vehicle, or at demographics where a social context and integration are more important than the physical characteristics of the car. The OEM values may be different across brands as well, so new scores will be required. The objective is to set a goal against which we can measure our investment or performance and to help steer decisions that need to be made during the specifications and design phase of a vehicle. The exercise we show through the generations is only an example against our goals for a cognitive life.

Let us go through our definition for each of self-enabling themes, then we will summarize the customer and OEM values for each generation.

Definition of Self-Enabling Values for the Customer

The value factors for the customer are associated with how much a customer is willing to pay and if the payments are distributed across different ecosystems.

1. Low value means that the services provided in the self-enabling theme are expected to be free of charges.
2. Medium value means that the customer is willing to pay some for the services in the theme, but the amounts are limited and the payments stay in one ecosystem.
3. High value means the customer is willing to pay higher amounts that may span different ecosystems.

We will now assign above value factors definition to the self-enabling themes for the cognitive vehicle:

Self-integrating   3, for helping the customers integrate with various things around them, which will create payments across multiple ecosystems.

Self-configuring   3, for adapting the vehicle to various customer wishes as they move through life, which will also create payments across multiple ecosystems.

Self-healing   1, for better convenience for the customer, which should be free services enabling standard vehicle maintenance.

Self-learning   2, for helping the customer to use personalized services that are based on the user behavior.

| Self-driving | 2, for safety and comfort, the customer is willing to pay for these vehicle functions. |
| Self-socializing | 1, for convenient services that should be free to the customer. |

Definition of Self-Enabling Values for the OEM

The value factors for the OEM are associated to how much the OEM can increase revenue or profit, or how much cost can be reduced.

1. Low value means that the services provided in the self-enabling theme are creating component level profit only.
2. Medium value means that the OEM is increasing revenue or is recovering costs for the services in the theme.
3. High value is creating new revenue from outside vehicle ecosystems or high cost reduction inside the vehicle life cycle.

We will now assign above value factors definition to the self-enabling themes for the cognitive vehicle:

Self-integrating	3, for creating new revenue for OEM from merchants' ecosystem.
Self-configuring	1, for more adaptable component profit only.
Self-healing	3, for primarily warranty cost reduction.
Self-learning	2, for increasing revenue by understanding and getting deeper insights about the customers.
Self-driving	1, for primarily sensor component profit only and cost recovery of providing automation services.
Self-socializing	2, for some revenue opportunities with Marketing and Sales, but mostly cost recovery.

3.1   Generation 0: The Non-connected Vehicle

To help understand value, we would like to re-introduce the automotive value chain from [26] as the foundation for the automotive industry we discussed in Sect. 1.1 of the introduction. The intent is to use and understand how the value chain is changing, who is doing the investing and who is gaining the value. Figure 3.1 shows the traditional or Generation 0 product and service value chains for the automotive industry.

The first buying point identifies the sale of the vehicle, in the case of automotive that buying point is at a dealership, usually not owned by the OEM. The second buying point identifies the sale of services, in the case of automotive again the buying point is at a dealership, usually not owned by the OEM. For Generation 0, the product and service value chain was driven but the physical nature of the AUTOmobile, the style, the brand, the horsepower, the cost and the features offered

Fig. 3.1 Product and services value chain

by the vehicle. Not many services existed after the sales of the vehicle, maybe some aftermarket options or maintenance plans offered by the dealer.

A vehicle in the US is considered the second largest purchase a person makes outside their home. For such a purchase, which happens more frequently than buying a house, it's important to be closer to the buying point. The history of the industry and the growth of automotive certainly show companies and products that have been successful, but will this continue? An interesting model since the OEM really doesn't have the relationship with the customer, it's really in the hands of the dealership. The value around customer experience and a personalized experience comes from being close to the customer and close to the buying point. Automotive in either case has not made the investment to be close to the customer's buying points, and hence is losing a lot of value and connection to the customer. Tesla and Apple are examples of companies that made the investments to get closer to the buying point with their physical products, hence gaining more value from the customer relationship. But that kind of relationship is not developed at an automotive dealership, because the person at the POS is only paid on the sale of the vehicle. As we mentioned in the introduction, is buying a vehicle a good customer experience, how many people like the relationship with their salesperson at a dealership?

These factors and more will be considered as we move through the generations and see how the value chain is changing and we see how what the industry is selling is changing as well.

3.1.1   External Ecosystem Changes

The physical components and ECUs drove the investments in Generation 0, with tier 1 suppliers tooling up to provide these components. OEMs retooled manufacturing facilities across the country, improved their supply chain and factory automation. Globalization brought more foreign manufactures into the US, Honda as the first Japanese automaker with a manufacturing plant in 1982, and BMW with the first German plant opened in 1994.

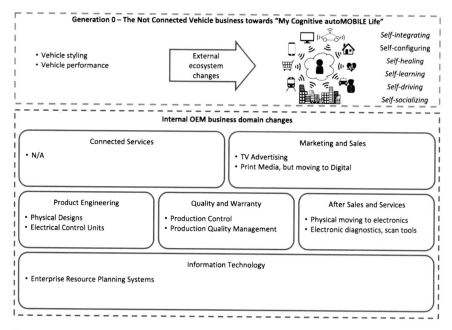

Fig. 3.2 Outside ecosystem changes impacting inside OEM business domain changes with Generation 0—non-connected vehicle

3.1.2 Internal OEM Business Domain Changes

The Generation 0 use cases and changing values of the OEM business domains were related to the vehicle styling and performance, (see Fig. 3.2), for a summary followed by a detailed breakdown for each department.

Connected Services
Connected services organizations didn't exist at this time.

Marketing and Sales
Traditional marketing and sales for Generation 0 used television advertisement, print media and dealership campaigns. The Internet was growing and online digital content grew related to the vehicle, what used to happen on the dealer lot could now be done sitting at home.

Product Engineering
Generation 0 value was in the physical world, customers purchased a vehicle for purpose, style, and horsepower, product engineering designed accordingly.

Quality and Warranty

The US auto industry knew it needed to improve. The Japanese OEMs were winning the quality races and really starting to affect US vehicle sales. Along with improving manufacturing on the physical side, IT systems helped monitor and track quality and drive the needed process improvements.

After-Sales and Service

ECUs are being used in the vehicle more, scanning tools and programming is popular at the dealership, technicians are moving away from mechanical skills into electronics skills. None of the ECUs had wireless communication capabilities, but the CAN bus, a physical network in the car was being developed as well.

Information Technology

'Standard Enterprise Resource Planning' systems started to make their way into automotive in the early 1990s. SAP was leading the pack. These systems improved finance, human resources, etc., but the fundamentals of supply chain management and just in time changed manufacturing and helped reduce costs in inventory.

3.1.3 How Changes Result in Value?

The ECUs in the vehicle were connected to each other, just not to the outside world. Embedded software was used in all the ECUs and in the early days couldn't be reprogrammed, so if a part was faulty, it had to be replaced even if it was just the software. Later in the 1990s and into Generation 1, reprogramming of ECUs was introduced, and managed at the dealerships.

Conclusions

The golden age of the AUTOmobile, physical styling and electronics features drove vehicle sales. Software and electronics were already starting to replace mechanical components in the vehicle and computer systems were helping the business and manufacturing to make better vehicles. Physical connectivity from the Internet, intranets, and the CAN bus drove the investments and the physical style and new types of marketing drove the vehicle sales.

Self-Enabling Values for Customer and OEM

We will review the self-enabling values for customer and OEM with Generation 0, (see Fig. 3.3). The only scoring that received any focus was self-configuring as OEM and product engineering started to invest in mechanical and electronic options that customers desired, but these only returned component profits to the OEM.

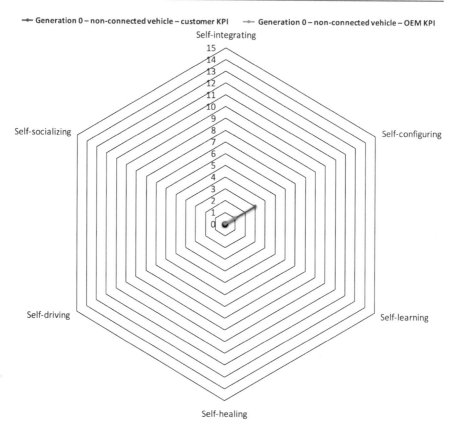

Fig. 3.3 Self-enabling values for customer and OEM with Generation 0—non-connected vehicle

3.2   Generation 1: Connected Vehicle

Moving into Generation 1, we now see the emergence of the second buying point in the Fig. 3.1, the connected vehicle buying point. These are the new services we introduced as Safety & Security, the key being they are sold after the purchase of the vehicle. Now look at the second buying point and the services around the vehicle. That same salesperson now has to talk to the customer about a monthly subscription that is worth a thousand times less as a single transaction than the value of the vehicle. Most OEMs did not invest in this case. The service providers like Verizon Telematics, SiriusXM, Wireless Car, etc. made the investments, owned the customer relationship and were closer to the buying point.

We are not saying OEMs didn't invest at all. Certainly, their engineering and business groups supported the functions and integrated them into the vehicle, but the core investment was made by the TSP. The result was not a good situation. Would Apple be as successful as it is today if it had tried to sell its smartphones and iTunes through third-party retail chains?

Not that the dealer model is broken, but selling services related to a major asset like a vehicle are difficult to provide. The industry tried automatic activations and free services for a few months or a year to incentivize the buyers related to the services. One difference to Apple was that services drove the value of the asset. For Apple, it was key to their business to own the asset and the digital platform. Where in the automotive industry, at that time, the value was in the asset and not the services. Another aspect that was difficult to manage was the renewals and new activations process after a vehicle was sold. The biggest problem definitely arises when the vehicle is sold a second time. Getting a cancelation and not knowing who the new buyer might be is not good for a subscription business. Sure, OEMs still had the branded websites that looked like they provided the service, but the reality was that the TSP gained all the value. The point isn't to assess the success or failure of the original TSP business model and whether anyone in the industry made money, but to point out investment versus value.

The value chain of Generation 1 services is detailed in the following five areas (see Fig. 3.4):

1. At a high level, first we have the tier 1 suppliers, including Panasonic, Harman and Alpine who were making the devices that are part of connected vehicle including the TCU and the head unit.

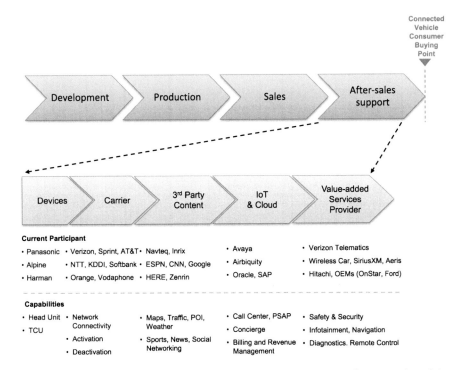

Fig. 3.4 Changes in the service value chain of the AUTOmobile with the first generation of the connected vehicle

2. The carriers in the US are Verizon and AT&T, which provide the wireless Internet connectivity.
3. The third-party content providers have core capabilities like maps, news, sports, music, weather, parking, etc. with companies like ESPN, CNN, The Weather Company, Parkopedia, SiriusXM and Pandora as examples. Some of the content providers and technologies related to the head unit weren't part of Generation 1, but their roles in the value chain was the same.
4. The IoT, cloud and service enablement are the core capabilities it takes to run the businesses and services companies are providing to automotive related to billing and revenue management, call centers, and public safety access points. These were the core capabilities owned by the TSPs using technologies from companies like Oracle, SAP, Avaya and Airbiquity.
5. Lastly, the value-added services or applications the industry has implemented over the years, which we will categorize into Safety & Security, Infotainment, Navigation, Remote Diagnostics and Remote Control.

Value-added services are what the end customer can see and buy, while all the others are really supporting capabilities. The TSPs invested and owned the value-added services at the time. Detailed definitions and descriptions of the TSP capabilities were described earlier in Chap. 2, and the alignment to the value chain will be used to understand and identify the areas where investments were made. Now aligning them to the value chain shows all the capabilities except the device were invested in and owned by the TSP. Device providers and third-party content providers made investments as well and being more of a component supplier provided something that had value.

3.2.1 External Ecosystem Changes

Analysis of Generation 1 and the value chain shows the first value-added services introduced to customers, their value in the case of Safety & Security was safety and comfort. If we ever meet anyone who has been in an accident in the US and had the emergency services automatically show up to help, we will understand customer experience, loyalty, and value. Nothing like having a customer demand for a service that the competition does not offer. That ultimately increases the sales of the product.

Most of the suppliers in the value chain are known as tier 1 suppliers and are defined as the major component suppliers to the OEM. The supplier value and business model can be quite competitive and an area where it is difficult to differentiate. The value of a tier 1 supplier is to bring technology and innovation to an OEM in hopes it will end up in a future vehicle. Once a vehicle is designed, performance specifications are sent out for a request for proposal and the tier 1 suppliers bid to win the contract. Tier 1 suppliers compete mostly on price, but with the correct technology and innovation, delivering that to an OEM as the best price will typically win. In the short term, this can drive revenue growth, but as a part/component becomes commonplace, it is more difficult to show the innovation, and cost becomes the driving factor.

The business model is simple: build to specification, control costs and warranty, create a profitable business. Early on in Generation 1, key device suppliers were the only ones in the market and the business model was good. The carriers started off as a commodity in our opinion and are still a commodity today. Cost and coverage were the key factors in selection, with little to no area for differentiation.

The suppliers of content had a much different business model, typically not driven by cost, but the basic capability to deliver the information required. Some unique content providers existed, but in general, it was the news, weather and sports providers. The suppliers for the IoT, cloud and the value-added services was the TSPs, and as we mentioned earlier, the companies that did most the investing and gained the most value. These companies had the opportunity to provide a unique service to the OEM and created the initial subscription models that drove the business.

In Fig. 3.5, the flow of money includes the customer, OEM and the TSP, where

1. first the customer buys the vehicle from the OEM or dealer
2. and then signs up and pays for the subscription through the TSP.
3. The TSP shares some revenue with the OEM
4. and pays the content and service providers.

The subscription-based model at the time was thought to be the only model that would be used in the industry. Until *freemiums* came along, we will go into that more in Generation 2, the industry's thinking was to generate the revenue based on the subscriptions. Subscription models aren't bad, SiriusXM has been very successful in

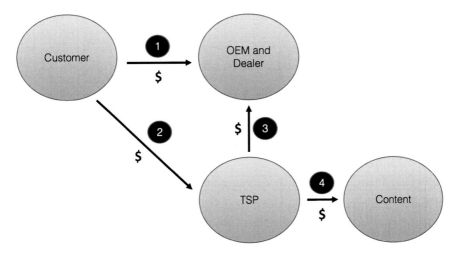

Fig. 3.5 Flow of money in a subscription-based business model

the industry with their satellite radio service. Netflix and other streaming providers have been successful, as well as cable companies, and the carriers for smartphones.

Initially, not much thought given to the internal value of the data, and not much data was collected, mostly due to cost. Neither was much thought put into the customer data or understanding the customer, and with the OEMs selling their services through dealers and TSPs, their connection to the buying point and the customer was lost. General Motors was the exception to the TSP model, being the only OEM to build their own infrastructure. At the time probably not much value, but as we will discuss later, the value would come.

3.2.2   Internal OEM Business Domain Changes

The Generation 1 use cases and changing values of the OEM business domains were related to Safety & Security, (see Fig. 3.6), for a summary followed by a detailed breakdown for each department.

Connected Services
This organization typically didn't exist during Generation 1, except for General Motors and OnStar. For the sake of discussion, we will not look at General Motors separately, but look at the typical overall market situation and organizational approaches from the OEMs.

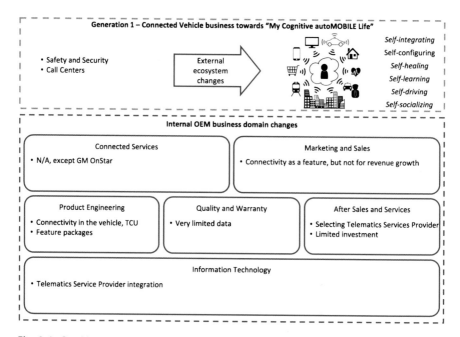

Fig. 3.6 Outside ecosystem changes impacting inside OEM business domain changes with Generation 1—connected vehicle

Marketing and Sales

Marketing and sales invested in campaigns that promoted the vehicle and new features related to the vehicle, ultimately looking to increase vehicle sales and differentiate their products from their competition. Early on, not much advertising investment was made in telematics features, the revenue compared to the vehicle was to low and at the time, it wasn't considered to be a factor in the sale of the vehicle itself. Initially, it was not easy to measure the increase in vehicle sales due to just the telematics features, so all that was tracked was who bought the option package and who renewed the service. For most companies, marketing and sales is considered a cost center and not responsible for making return on their investment other than promoting the brand value and growing product sales.

Product Engineering

The product engineering organization made the investment to integrate the TCU into the vehicle, select the supplier, and as for most parts of a vehicle, their business model or measurement was to recover the cost of the part in the final price of the vehicle. It is actually a combination of product engineering, marketing and sales that selects the features in a vehicle best suited for that vehicle's brand and demographic. Vehicle features or options were sometimes the factors that drove vehicle sales, so an important process was not just to put features in a vehicle, but also to understand the customers and their inclination to buy those features.

At the highest level, product engineering is measured on the sale of the product and the sale of the option packages, both easy to track and understand their return on investment. Early on, most telematics features were part of option packages, so it was easy to track the success of the feature. As we move through the generations, more and more of the features will be part of the base vehicle package, making it harder to measure the direct ROI.

Quality and Warranty

For Generation 1, the quality and warranty organization didn't make much investment because the telematics service didn't send a lot of data from the vehicle due to the high cost in wireless transmission fees.

After-Sales and Service

The OEMs did not initially setup a connected services organization. We will expand on that later. The after-sales and support organization typically owned the responsibility for the overall success, both profit and loss of the telematics services. This organization covered the investment or selection of the TSP. Not much investment was actually needed, it was more about selecting the appropriate supplier to provide the services, and getting part of the revenue from the monthly subscription. After-Sales and support worked with the dealers on how to promote, sell, and support telematics services, for both new and used car buyers. Interesting was that the OEM or their TSP was responsible for the renewal process, not the dealer.

For the overall return of investment, the cost of the after-sales and support team plus the IT costs were measured against the revenue generated by the telematics service and the overall value being an increase in revenue, outside the sale of the vehicle. Since we never worked for an OEM or have seen their detailed financials, we will not comment on the results of the business case specific to any OEMs, only to point out how the business model worked.

Information Technology
The IT organization made investments to integrate the TSP into their business because key information was needed by the TSP to supply the services, such as what vehicles had a TCU and who bought a specific vehicle. IT also helped integrate the TSP websites into the OEM's customer websites, so if the service wasn't purchased at the dealer or the customer was a second buyer, customers could sign up or renew services on the web. Typically, these investments were considered a cost in the overall business case and had to be recovered or offset from the revenue generated by the telematics service.

3.2.3 How Changes Result in Value?

For Generation 1, the main challenge was how to grow a services revenue model and make sure the features offered to the customer were valuable enough to do so. The first problem was sign-ups and renewals. Whether the OEM offered a free trial period or the dealer sold the services directly, this was difficult initially because the salesperson that sold vehicles was now selling services at a fraction of the cost or reward. With the free trial, this responsibility moved away from the dealer and into the hands of the TSP, who was contracted by the OEM. Both had an interest in signing up and renewing customers, which was the main driver of revenue. Early on, one of the main problems with Safety & Security was that it was really just a type of insurance policy. Must customer didn't get a chance to use it or see the benefit, but if the customer was in an accident or helped after a breakdown, the service did sell.

Conclusions
As we continue to analyze the generations, the changes and decisions related to the organization become key factors in the products success. This is not unique to automotive, but any product-driven company looking to transform into a services-oriented company. As other industries look at their value chains, understand buying points, understand the connection with the customer, understand where they need to invest and why, they can learn from what automotive did early on in their attempts to become service-oriented companies.

Self-Enabling Values for Customer and OEM
We will review the self-enabling values for customers and OEMs with Generation 1, (see Fig. 3.7), starting with customers seeing more value in self-configuring and self-integrating, as the OEMs and after-sales and service invested in new connected vehicle services, which also brought them new sources of revenue as well. The

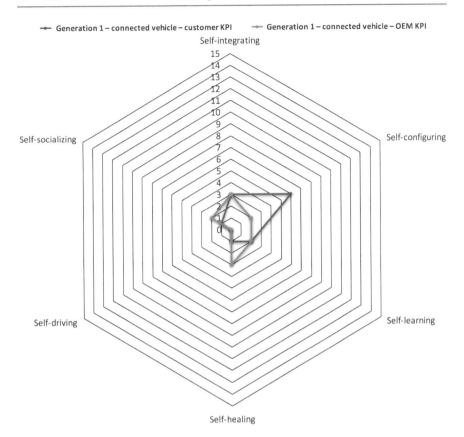

Fig. 3.7 Self-enabling values for customer and OEM with Generation 1—connected vehicle

OEMs also started to see the value of monitoring the vehicle and getting the first look at diagnostics events from the vehicle.

3.3   Generation 2: Infotainment and Navigation

The value and business models really started to change and transform in Generation 2. We didn't see much change in the buying points. New value-added services were introduced, but the problem with dealerships and selling services continued to be a problem. The automotive market certainly did not keep pace with the smartphone and app store revolution, but the unique nature of providing these types of services while driving provided a challenge for the industry. The introduction of the head unit in the vehicle for entertainment and navigation purposes brought more suppliers and capabilities to the vehicle, but as smartphones became popular, people started to question the value of these features being included in the vehicle. Some OEMs took a route that focused more on the value of the smartphone integration to the vehicle rather than integrating functions into the vehicle. This led OEMs to take one of three strategies built-in, brought-in or beamed-in as the platform in the vehicle to provide the value to their customer. We will use the following definitions for these terms:

1. A *built-in* platform involved both a TCU and a head unit with full connectivity to the Internet. The applications were developed and deployed on the head unit, using the connectivity to get the content or data they required. Classified as a high-end solution, the cost of the monthly connection and the high cost for embedded application development created difficulties in the business model, but it was still used by most luxury brands.
2. A *beamed-in* platform involved a head unit similar to a built-in platform, but with more of a focus on applications being in the cloud and just the connectivity provided by the smartphone. Since the monthly connection cost was an issue, using the customer's connectivity and data plan was one way around the problem, but certain scenarios like remote door unlocking or automated accident notification were not possible.
3. A *brought-in* platform involved a head unit with limited functions that integrated with a smartphone for the applications and connectivity. Ford was the first OEM to adopt this strategy with Ford Sync and its AppLink platform, but now both Apple CarPlay and Android Auto provide a similar functionality.

Hybrid combinations also emerged as Apple CarPlay and Android Auto became very popular for certain functions in the vehicle and could easily be included in built-in or beamed-in platforms.

Why so many options? Why not just let the smartphone industry take over these functions in a vehicle? Well, it all comes down to value and the customer experience. As the value-added services started to increase in the vehicle, it was no longer just the TSP or OEM that provided these functions. New players entered the market, either in the vehicle or on the smartphone. A more detailed picture of the value chains shows the increase in value-added service related to Generation 2, (see Fig. 3.8).

This started the debate on which way was better, and which way provided a better customer experience. Take the simple example of navigation and traffic information. Originally launched in vehicles before the smartphone, customers could really start to see the value of not just having a navigation system, but also getting real-time traffic information. Honda with AcuraLink in 2005 was the first system to update traffic in a vehicle over the air through SiriusXM, which at that time provided the best value and customer experience. The business model was subscription based through SiriusXM, with the ability to add the traffic feature along with other premium audio channels. Then in 2009, Google announced Google Maps with traffic on the smartphone for free. Now the debate started which way was better. Of course it wasn't just about which approach was better, the main argument from OEMs was not wanting to put the smartphone industry at the forefront of the customer experience in the vehicle.

Think back to Chap. 2 and the lessons learned related to Generation 2. The HMI, customer experience and now value, all start to play a key role in these decisions

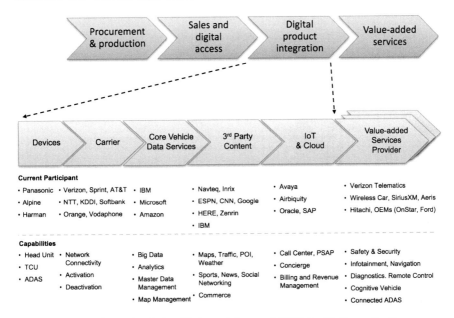

Fig. 3.8 New services value chain with the evolving autoMOBILE with Generation 2

and options. Let us look again at our simple example of navigation and traffic. Great value for the customer, but with a bad HMI and customer experience, what value is the customer really gaining? Other new value-added services in Generation 2 did provide value combined with a good customer experience. Bluetooth hands-free provided a safer driving experience while talking on the phone, other Bluetooth technologies and smartphone integration allowed the customer to bring his own music and applications into the vehicle in a more seamless fashion. Dynamic POIs and weather were also integrated into the navigation system, providing additional value to the driver to navigate related to these types of changing situations. To summarize the customer value in Generation 2, it was mostly focused on convenience, reduced driver distraction, and entertainment.

3.3.1   External Ecosystem Changes

Tier 1 suppliers, TSPs, and content providers did not really see much change in Generation 2 other than increased growth in vehicle adoption, which certainly translated to their bottom line. A few content providers like TomTom, Inrix, and SiriusXM grew their business of bringing dynamic content to the vehicle, using a subscription-based model that customers purchased through them directly. The *freemium* business model introduced in the smartphone industry certainly started to put pressure on the subscription model used in context of the connected vehicle. Of course, most people think freemium means free, but the reality is that the consumer

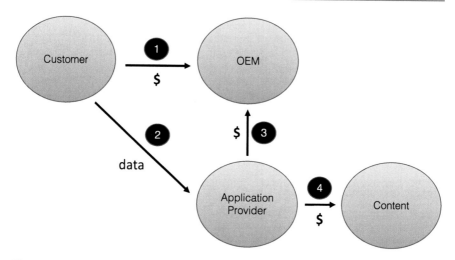

Fig. 3.9 Flow of money or data in a freemium business model

are typically giving up something to get that application for free, either the location, personal information, or dealing with advertising.

Figure 3.9 looks similar to a subscription model (see Fig. 3.5) except the provider is not collecting money from the customer in 2. It collects some type of information in return for providing a service for free.

The freemium model certainly pushed the debate on what type of solution was best for the vehicle, customers commented:

▶   "Why should I pay for it in my car when I can get it free on my phone?"

A good argument and maybe why some OEMs are going more with the brought-in strategy, but if we look at the overall problem, we believe the main factor is driver distraction. Before Apple CarPlay, if a customer wanted an application with a rich user experience and designed for limited driver distraction, it had to be built into the vehicle as a native application. Sounds easy, but head units are not like smartphones and at the time most of the platforms were unique, so it was expense for the application providers to support such a wide range of platforms. The industry certainly thought there would be as many apps for vehicles as we have on our smartphones. Well it didn't happen that way, due mostly to cost and the introduction of technologies similar to Apple CarPlay that eventually made it easier for application providers to get their apps in the vehicle. So is built-in completely going away? Probably not. As we will see more in Generation 3 and beyond, certain types of functions related to navigation and safety will always require dedicated systems in the vehicle, but it is safe to say the basic entertainment and smartphone integration will still exist as well.

The main transformation besides the additional value-added services is the OEMs looking to control more parts of the value chain as more data from the vehicle and about the customer are available. One major change to the value chain is the addition of what we will call "core vehicle data services" (see Fig. 3.8). By definition, as the vehicle is now able to create and transmit more data, how to collect, store and manage these data becomes a key capability. OEMs are now investing in this area to gain core value from the data to drive mostly internal value, warranty cost reduction, predictive maintenance and a better understanding of their customer and their behaviors.

3.3.2 Internal OEM Business Domain Changes

The Generation 2 use cases and changing values of the OEM business domains were related to infotainment and navigation, see Fig. 3.10 for a summary followed by a detailed breakdown for each department.

Connected Services
In Generation 2, OEMs started to reorganize centered on a connected services organization that was responsible for the strategy and deployment of these new connected services. These were typically made up of people from a cross-section of the company including especially IT, product engineering, after-sales and service and marketing. Their aim is to improve communication and decision making related to connected services. Their biggest budget item was IT and supporting the other

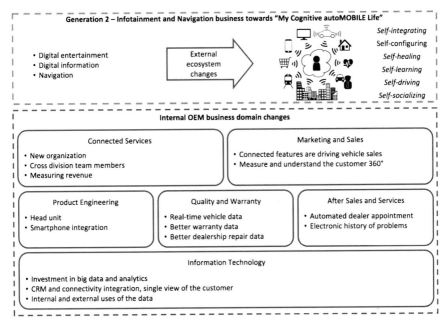

Fig. 3.10 Outside ecosystem changes impacting inside OEM business domain changes with Generation 2—infotainment and navigation

organizations, who gained most of the value, in soft terms, by not making a lot of investment. The revenue from external data did help with the overall ROI and bring that into the control of a new organization.

Marketing and Sales
Probably the most affected organization in Generation 2 as far as changing their thinking and approach was concerned was marketing and sales. Previously, the telematics features were basically ignored and the revenue wasn't important. Now telematics, navigation and infotainment were important factors in the customer decision to buy a certain type of vehicle. This meant advertising these features, and understanding how to measure the effects of base package features on vehicle sales. Great news for the industry and customers as far as it concerns adoption, but difficult to measure for marketing and sales.

Product Engineering
Product engineering continued to invest, bringing additional hardware like the head unit to the vehicle along with smartphone integration and entertainment functions. Similar to Generation 1, the telematics feature cost was recovered in the price of the vehicle and the organization was measured in terms of sales of those telematics features. We did see in Generation 2 the emergence of the telematics features driving vehicle sales, i.e. people wanted vehicles with these types of features and that drove vehicle sales. One simple example was smartphone integration. Vehicles without it suffered in sales to the point where now all vehicles have some type of smartphone integration.

Quality and Warranty
Quality and warranty did not have to change their thinking. They had to adopt to the fact they were starting to get real-time information from vehicles. This data along with better data from their warranty and dealer systems helped reduce warranty costs, possible shorten the lead time for a recall, and improve vehicle quality. The investment was mostly related to IT with technologies for big data and analytics. The ROI was a bit harder to measure since it was difficult to establish a one-to-one relationship of a warranty or quality issue to collecting real-time data. Probably one of the largest internally valued items, but the most difficult to measure.

After-Sales and Service
The after-sales and service organizations slowly transition key responsibilities to the connected services organization. At the same time, after-sales and service also benefited from the vehicle data. Diagnostics codes and the history of codes helped service technicians diagnose problems at the dealership, leading to improvements in customer satisfaction and the ability for the dealers to correct the problem on the first visit. These are examples where it is hard to measure the value of the solution when it is directly related to customer satisfaction.

Information Technology

IT is now supporting all the data collected from the vehicle and the multiple organizations that want to use it. Both the marketing and sales and the quality and warranty organizations are using big data and analytics tools not just to learn about the vehicle, but to start to learn more about the driver as well. The connection to CRM and the single view of the customer is a complete picture when real-time data from the vehicle and driver are used. Data are not used just internally. Some OEMs start selling data externally. Partnerships could also be established with external companies in the area of PAYD or PHYD insurance, who used vehicle and driver data to provide benefits for the driver in the form of reduced insurance cost and for the insurance companies in the form of reduced risk. Location data from GPS can be used to improve traffic services and even control toll collection in cities. These external services are another source of revenue along with the monthly subscriptions. Value also exists in the vehicle. Some OEMs are creating frameworks for access to the data inside the vehicle for applications as well. OEMs that are the early movers in this field will be well positioned to provide new value-added services and gain the value from the data. Along with the data, IT has picked up development and support for the mobile applications used for remote functions and vehicle statuses. Still the ROI is harder to measure, more internal organizations are using the data, but some new revenue sources have been developed.

3.3.3   How Changes Result in Value?

In most companies, sharing cost and revenue for a product is difficult when those factors span so many different organizations. Traditionally, a product-focused company managed their product cost in product engineering with a basic model that required the cost of the component to ultimately be covered in the price. Other organizations tend to be cost centers and along with manufacturing add up to determine if a product makes or loses money. This is a simple business view of a company, but as we have already seen in Generation 1, the IT organization is not just a cost center, but clearly part of the cost of the product and its ongoing cost to support the functionality. Value within companies also does not necessarily come from the ones who made the investment. Why would an engineering department design and support collecting data for warranty purposes when that is not their department? Marketing always wants more information about the customer or the product usage, but we have seen systems built that could have easily supported this, yet did not. So some of the hype around the IoT is valid and can be seen even in automotive as the industry matures and moves from a connected vehicle to a cognitive vehicle. Not only do costs start to become a factor, but decision-making and design are more difficult as well.

Let us talk briefly about some of these different development approaches and how people throughout the years have been changing them, starting with product engineering, whose development process was based on the V-Model approach. The advantages of this approach are the focus on performance specifications and quality, understandable when building real-time systems to control an engine. IT started out with the waterfall model [3], similar to V-Model except testing was not done as much in parallel to the design. IT has been changing more and more over the years with the introduction of agile methods, where an incremental and iterative process shows progress along the way, (see Sect. 1.4). Then Gartner came along and started to discuss bimodal where IT organizations should use a blend of approaches driven by demands for digital and IT to be more responsive, see [5]. Design thinking, which we defined in Sect. 1.4, was also introduced, but it's not a development process, rather a better way to capture requirements and understand the customer, which is also call user-centric design. We do not want to debate these different approaches or understand, which one is best to use when, but focus more on the point that all these organizations have different backgrounds, which will cause problems. For a more in-depth look at the software process evolution, see the literature [2].

Organizations that typically do not work together are now all part of product requirements and design. Common problems start with engineering and product development companies that typically do not work with smartphone technologies, Internet technologies and cloud technologies. The IT organizations also have problems because they don't understand the technology behind complex embedded systems. Software engineers usually coded their own operating systems and had real-time control of their program inputs and outputs. This is not how the Internet and the open-source paradigm work today. IT organizations had to start working with engineering departments and they realized they had completely different development processes. Then throw in marketing and other organization that want to have their input on requirements and features on a daily basis where engineers are used to working to a fixed specification. We have even seen examples where there has been such disconnect between organizations that requirements are being specified for a product that doesn't even have the ability to perform them. We have even heard of cases where the data is available, but no one is using it, so the tools and processes around making data available to the internal organizations is not without its own challenges. We can typically tell this is a problem when a connected services organization asks about creating more value and revenue related to their vehicle data, but they can't even answer basic questions around their connected service business, i.e. how it helped warranty or marketing. Remember, value comes in many forms, additional revenue is the easiest to track, but do not ignore the rest.

Conclusions

Like any strategy, most OEMs will choose their own path, some focusing on the value of the data and connected services, others focusing on the traditional automotive features. These are just some of the many examples of the value of data from a vehicle. The overall metrics for connected services could also be better managed by a single organization, but still some of the soft value was difficult to measure across the other organizations.

One thing other industries can learn as they asses their value chain looking at automotive is no longer focusing on a cool technology, i.e. to unlock q car door with a smartphone, but focusing on value-added services that bring the OEM closer to their customer and promote their brand. White goods are a perfect example. Mostly sold through major retailers in the US, OEMs are not connected to the buying point, and the only way they get customer information is through the warranty card. When was the last time we filled out one of those? Connected appliances, the value-added services, were the first instance that brought the OEM closer to the buying point and closer to the customer. The value these OEMs are gaining is very similar to automotive and another example of connectivity affecting the business.

Beyond the OEMs, the suppliers are evolving into two types of companies. On the one hand, large system suppliers like Bosch which are close partners of the car manufactures but also adapting IoT business outside automotive, and on the other companies that are going out of business because they only focused on specialized components for a single industry and did not adapt to the changes required by digitization.

Self-Enabling Values for Customer and OEM

We will review the self-enabling values for customer and OEM with Generation 2, (see Fig. 3.11) with the biggest change for the customer related to self-configuring

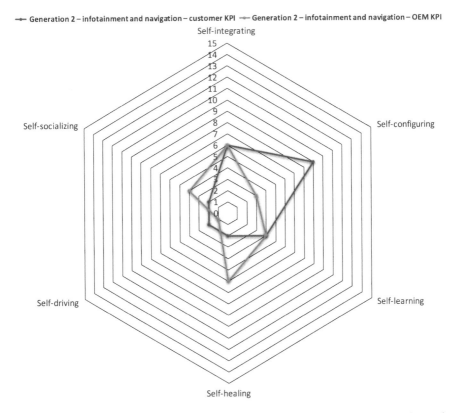

Fig. 3.11 Self-enabling values for customer and OEM with Generation 2—navigation and infotainment

and self-integrating based on the OEM investments in infotainment. Product engineering and the new connected services organizations were bringing a lot of digital options to the customer in the form of digital entertainment or information. With the investment in connectivity, quality and warranty started to see their value increase with warranty costs reductions.

3.4   Generation 3: Cognitive Vehicle (Today)

For Generation 3, the focus on customer experience and the use of vehicle data is becoming key, and the phrase "data is the new oil", which is actually more than 10 years old, is getting popular in automotive.[1]

Continuing with the growth of value-added services from Generation 2, OEMs learned the value of being closer to the buying point and building a stronger relationship with their customer from other industries. The cognitive vehicle is an area most OEMs are investing in. They realize now the value of owning the personalized interaction with the customer, not just because what can be learned about the customers, but ultimately how that relates to the brand identity. These types of new value-added services will be a clear battleground in the future. People always ask why do we need a cognitive vehicle and CPA when we have Apple CarPlay and Android Auto?

Certainly some OEM will give away the personalized interaction to the mobile industry, but that will only go so far in the vehicle. Look at the concepts for the future digital cockpit and experience in the vehicle, can that be realized just by plugging in the smartphone? No, there certainly is a place for some basic navigation functions and personalized entertainment, but the driving experience will be owned by the OEM.

Recently, lots of announcements were made related to the CPA, for example on the Consumer Electronics Show in 2017. But were they really a personal assistant. As we defined earlier, a CPA as something that learns, adding a voice interface upstream of existing functions like unlocking the car door from the house, is not a CPA. Could they grow into that? Sure, but the first generation of assistants are simple just that. We don't see yet the type of investment to gain real value from this, but OEMs are surely looking at it.

As JD Powers shows, the existing voice recognition systems in vehicles leave a lot to be desired.[2] Having to speak defined commands and only at certain times isn't a very good customer experience. The problem is the lack of NLP implemented in the vehicles and the cost to do so. Many new capabilities are available through the

---

[1]"ANA Marketing Maestros: Data is the New Oil" http://ana.blogs.com/maestros/2006/11/data_is_the_new.html. Accessed: July 26, 2017.

[2]"Bluetooth Connectivity, Voice Recognition among Top 10 Problems in 3-Year-Old Vehicles"   http://www.jdpower.com/cars/articles/jd-power-studies/bluetooth-connectivity-voice-recognition-among-top-10-problems-3-year. Accessed: July 26, 2017.

cloud, but automotive is unique in that it must deal with delivering features while being connected to the network and while disconnected from it.

The future CPA, with the correct investment from OEMs, will need to address the in-vehicle issue, but also create a personalized experience through learning that can assist the driver with their vehicle or their driving experience. BMW announced their CPA concept as the companion.[3] The value to the OEM from a CPA is easy to understand. Can we think of a better way to learn about our customer? We can't think of many. Of course, customers will want something in return. They don't give personal information away for free, but if it creates a unique and personal driving experience, it gets our vote.

3.4.1   External Ecosystem Changes

In Generation 3, OEMs started to look beyond applications and looked at other industries as well to understand how advertising and the business models for social and smartphone apps could apply to the vehicle. IBM and General Motors were the first to make announcements about introducing merchants into the vehicle.[4] The best example is the 'pay at the pump' system of Exxon, not just an advertisement, but an interaction in the time of need.

Google has also made a similar announcement related to its navigation application, driven by the ability to prioritize merchants in searches as well as being able to identify them on the map.[5] Merchants see value in knowing exactly where someone is so they can better target an advertisement or coupon. Other interests arise when promoting a brand like Starbucks, using vehicle or weather data to suggest a warm or cold cup of coffee.

Aligning brand and loyalty across the ecosystem and with the OEM is the beginning of the foundation of a connected life. The customers don't just want to see all types of POIs on the IVI, but only the merchants they are loyal to. In the future, OEMs may choose to align with other brands and use the car key or a token to identify individual customers, not just when they are in the car, but when they enter a hotel or even restaurants. But as OEMs also enhance the types of POIs, the POI or merchant can also become an interaction so now the customer doesn't just know where the restaurant is, but can place an order, pay for it and pick it up. Now that's customer experience.

We personally like the idea of paying for gas before we get out of the vehicle or being reminded that our tank is low and to stop at our favorite station on the way

---

[3]"Companion" https://www.next100.bmw/en/topics/articles/companion.html. Accessed: July 26, 2017.

[4]"Hello, OnStar – Meet Watson" https://www-03.ibm.com/press/us/en/pressrelease/50838.wss. Accessed: July 26, 2017.

[5]"Advertise on Google Maps with Google AdWords" https://adwords.google.com/home/resources/advertise-on-google-maps.html. Accessed: July 26, 2017.

home. What about a notification for a traffic accident and an automatic rerouting instead of picking up the smartphone and checking for traffic indicators? Weather notifications are another example. Maybe now is a better time to stop for lunch so as not to drive in the rain. Finding a parking spot is another example and in crowded cities, this can really be of great value to a customer in terms of saving time, when it is better to reserve and pay for a spot while driving into the city. These examples are technically not challenging, but clearly show simple examples of customer experience, and experiences that are related to driving and provide a bit of *e-motion* as well.

Content and interactions with merchants or a merchant ecosystems is an example where the OEMs are investing and gaining value from shared revenue models, similar to those that have driven the likes of Facebook, Google, and others. In Fig. 3.12, the flow of money is described as follows:

1. We start with the customer purchasing the vehicle from the OEM, but really through the dealer.
2. The OEM has paid a tier 1 supplier to build a head unit with these features, which the customer most likely purchased as an option package.
3. The customer uses a merchant service to pay for gas at the pump.
4. The tier 1 integrates the head unit with a service provider who integrates the content of the paying merchants.
5. The revenue collection from the merchants is the foundation for the model.
6. The rest of the model explains how the revenue is shared, which is similar to any smartphone advertising business model. The service provider shares the advertising and interaction revenue with the OEM.
7. The service provider shares the revenue with the tier 1 supplier as well, depending on who made the investment to bring the solution to market.
8. The service provider is bringing new merchants on board the ecosystem.

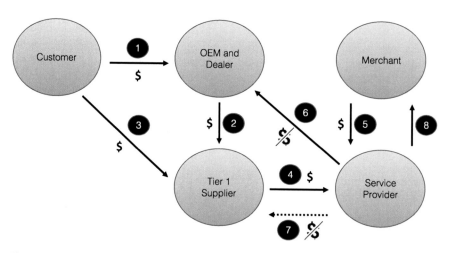

Fig. 3.12 Flow of money in a revenue sharing business model with options (dotted line)

Google has a similar model, but doesn't share the revenue with hardware suppliers in the smartphone industry, same for the smartphone application providers as well.

Adding to the merchant ecosystem, let us look at a few external examples starting with the recent announcement from BMW to provide vehicle data in an open framework to third-party developers.[6] Ford also announced in its mobility experiments the intent for third parties to develop applications based on vehicle data.[7] Ford[8] also announced an investment in Pivotal, a cloud software company, to support development activities related to its mobility experiments and even open sourced its in-vehicle application technology Ford AppLink[9] to promote access to data in the vehicle as well. General Motors has also made data available for applications both inside and outside the vehicle and promotes to third parties through is developer cloud.[10]

3.4.2 Internal OEM Business Domain Changes

The Generation 3 use cases and changing values of the OEM business domains were related to the CPA and the merchant ecosystem in the vehicle, see Fig. 3.13 for a summary followed by a detailed breakdown for each department.

Connected Services

New business models are getting relevant to grow revenue outside of selling services and subscriptions in the merchant ecosystem. Connectivity is gaining popularity around the globe, so investments in global solutions with a regional touch are new areas of investment. This strategy of bringing the data in-house puts less a focus on the TSP and in some cases relegates those providers strictly to providing Safety & Security services. Other OEM strategies and decisions related to selecting or controlling other parts of the value chain come into play as well, related to global carrier contracts and value-added services that mostly affect the customer experience.

---

[6]"BMW CARDATA. Your vehicle data. Permanently under control." https://www.bmw.com/en/topics/fascination-bmw/connected-drive/bmw-cardata.html Accessed: July 26, 2017.

[7]"Ford at CES Announces Smart Mobility Plan and 25 Global Experiments Designed to Change the Way the World Moves" https://media.ford.com/content/fordmedia/fna/us/en/news/2015/01/06/ford-at-ces-announces-smart-mobility-plan.html Accessed: July 26, 2017.

[8]"Ford Invests in Pivotal to Accelerate Cloud-Based Software Development; New Labs Drive Ford Smart Mobility Innovation" https://media.ford.com/content/fordmedia/fna/us/en/news/2016/05/05/ford-invests-in-pivotal.html Accessed: July 26, 2017.

[9]"Ford and Toyota Just Became Allies in an Unusual Agreement" http://fortune.com/2017/01/04/ford-and-toyota-just-became-allies-in-an-unusual-agreement/ Accessed: July 26, 2017.

[10]"Welcome to our developer site. This is where you learn ways for your applications to interact with our vehicles. Find inspiration in our gallery, or explore our APIs and begin innovating today." http://developer.gm.com Accessed: July 26, 2017.

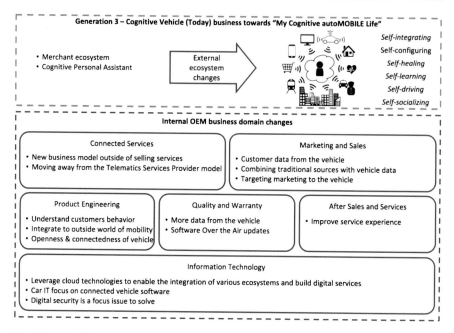

Fig. 3.13 Outside ecosystem changes impacting inside OEM business domain changes with Generation 3—cognitive vehicle (today)

Marketing and Sales

The investments for understanding the customer are an important value and combined with the traditional sources of data from marketing campaigns will improve the product and ultimately vehicle sales in the future. The organization is also getting used to the change where features are now software based and not just mechanical.

Product Engineering

With the drive and investment for understanding the customer, product engineering for the first time was able to see real customer usage information, not just data from after-the-fact studies. This knowledge could be used to improve the product and better understand customers' behaviors. Integration of this information into design processes will take time, but the value related to continuous improvement is clear. More investment in software development will happen to enable CPAs and merchant ecosystems, which will not bring direct revenue for the products teams, but for the connected services organization. These investments will challenge the vehicle design to consider various scenarios from the outside world of mobility. This drives more openness and connectivity requirements for the vehicle development.

Quality and Warranty

The quality and warranty organization continued to gain value in Generation 3. More and more data were collected from vehicles, which meant more improvements in quality and reductions in warranty cost. With the ability to update software OTA, some quality issues could now be resolved without a trip to the dealership. Even simple updates of the maps no longer required trips to the dealer, and some OEMs are even starting to sell vehicle features OTA, where it's much easier to release changes in the future as well. Regulations how to conduct recalls in an age of remote fixes and over-the-air updates will change.[11]

After-Sales and Service
After-sales and service continue to gain value from connectivity. Dealers can also join merchant ecosystems to build relationships with the customer to improve service and schedule dealer appointments. However, still mostly physical and human-based in a growing digitized world.

Information Technology
IT continues to be at the center for investments related to all the new digital services. More data and more users of the data continue the move to bring as much of the data in-house as possible. With the customer focus CRM, integration with merchants and the vehicle put a new perspective on the personal profiles and managing the person from the OEM's viewpoint. IT is leveraging cloud technologies to enable the integration of various ecosystems and build digital services. Digital security is getting more and more relevant, but reliable solutions do not exist.

Newly established "Car IT" organizations focusing on connected vehicle software are moving more into product engineering related organizations and traditional vehicle design processes. That does not provide IT the needed openness towards new business models, remember Conway's law (see p. 74):

> organizations which design systems ... are constrained to produce designs which are copies of the communication structures of these organizations

As a result, the OEMs want to build their own ecosystems around the AUTOmobile (see Fig. 3.14) [4], which is different than the real target to achieve 'My Cognitive autoMOBILE Life' (see Fig. 2.30).

3.4.3   How Changes Result in Value?

Of course, these new strategies and transformation don't come without challenges. Security and trust are becoming key factors in people's decisions on how companies use their data. Sure, most smartphone applications are using data and people are aware of it, but if they do not know or trust the brand using the data, the likelihood of

---

[11]"Over-the-air updates may alter NHTSA recall policy" http://www.autonews.com/article/20170123/OEM11/301239815/over-the-air-updates-may-alter-nhtsa-recall-policy.   Accessed: July 26, 2017.

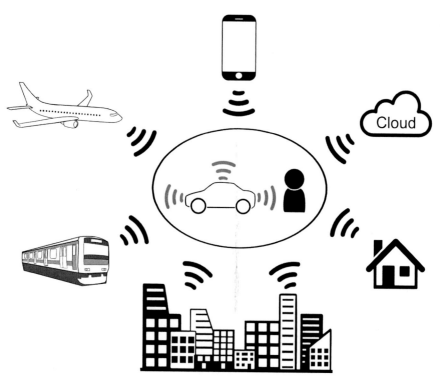

Fig. 3.14 In Generation 3, OEMs want to build their own ecosystems around the AUTOmobile, which is different than the real goal to achieve 'My Cognitive autoMOBILE Life' (see Fig. 2.30)

approving the application decreases. Smartphones also pioneered the transformation around freemium we discussed earlier, but the key to that business model is the customer gaining the value back. To add to the complexity, customers' behavior is different across the global regions, so systems need to be flexible in how they support their customers.

Conclusions

In Generation 3, we start to see the change from AUTOmobile to autoMOBILE. When we look again at the automotive value chain (see Fig. 1.3), in yellow we are starting to see the shift from development and production (focus on the AUTO) to digital product integration and value-added services (focus on the MOBILE). The merchant ecosystem, CPA, OTA and value-added services related to the data show progress towards this transformation. It's good to see some OEMs entering the platform business, BMW, Ford, and General Motors. This will certainly help them better understand the market and players as they move into the mobility industry. Some of the issues they may encounter on this journey include getting other brands to participate, how to deal with standards, and how to deal with the partners. Other brands will certainly not want to develop to each OEM specification, making it difficult to establish participation. A good reason to possibly see standards in this area in the future and understand how to collaborate with the smartphone industry to make it happen. Of course, automotive could follow the home automation industry, where we basically have three de-facto standards Apple, Google, and Amazon.

Remember that in [7] other industries can also look at the organizational challenges and changes that automotive has had to make with the cognitive vehicle. The real challenge lies in whether the directions with Generation 3 are leading to the right long-term differentiations for the next-generation cognitive life. IT might be the only organization able to set the technology strategy for the corporation. The organization can get stuck in its own ecosystem around the autoMOBILE, which will not grow fast enough and will limit the cognitive life demands (see Fig. 3.14).

Self-Enabling Values for Customer and OEM
We will review the self-enabling values for customer and OEM with Generation 3 (today), (see Fig. 3.15). Again, the customer is seeing value related to self-configuring and self-integrating as the CPA and smartphones are becoming

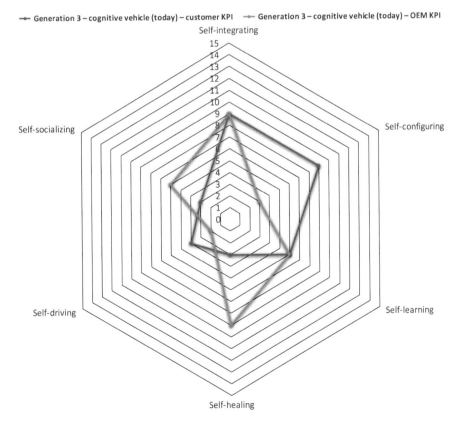

Fig. 3.15 Self-enabling values for customer and OEM with Generation 3—cognitive vehicle (today)

seamlessly integrated into the vehicle. The new revenue from the merchant ecosystems brings value to both the customer in the form of more personalized services, and to the OEM as an additional channel for revenue. The connected services team is making the correct investment, helping to bring those new services to their customer. Safety-conscious customers are also buying the new ADAS features and are getting a glimpse of more self-driving features to come. Product engineering is aligned with those investments as well.

3.5   Generation 3: Cognitive Vehicle (2022)

In Generation 3 (2022), we can imagine more mature value chains and business models taking hold without much change. For automotive to start realizing a personalized experience, the OEMs see the need to focus on the cognitive edge, partly taking the cloud aspect of the CPA and bringing it into the vehicle. Not only does the CPA need to move to the edge, but the identity and personal information need to be managed and synchronized with the cloud as well. How else can we achieve scenarios where we share the identity between vehicles or we update our vehicle after our kids change all the settings? Imagine configuring the screen layout in the vehicle with our own digital persona creating a look and feel that is much more personal. Everyone knows the value of creating a personalized experience, who would use a smartphone if we could not personalize it by downloading applications? We also see this type of value tied to brands. This is where customer experience and a personalized experience overlap, and once we have established this association, sales will certainly benefit. Customers again will see the value of safety coming back into focus, and now the need for dispatching emergencies services will decrease as ADAS and semi-autonomous features keep drivers safer. Maybe insurance companies can enter the business model with all the savings from fewer accidents, and with the addition of the connected data maybe it is even safer. Certainly scenarios where the vehicles are outside the range of the embedded sensors, highway situations at higher speeds, accidents ahead, and vehicles around the corner. This could be the case to justify the savings for sharing vehicle data in the cloud. Whether there will be enough differentiated scenarios to drive that kind of saving, we will have to wait and see.

3.5.1   External Ecosystem Changes

Another interesting area where OEMs are investing is related to content providers and the investment BMW, Mercedes and Audi made in their acquisition of HERE maps. Remember, it was as recently as 2009 when Motorola announced the first Droid smartphone, and Google Maps was a free application. The next day, the stock of Garmin, TomTom, etc. lost value in the stock market[12] and everyone said the IVI industry was done. Well, the IVI market is still growing, map companies have recovered and more options are available for displaying maps and directions inside the vehicle. Why such a big change of events? One reason is simple, maps will play a clear role in autonomous driving in the future, the other is that the new value-added services that can be provided to users around POIs.

Most people think of POIs as places, food, coffee, hotels, parking etc., as we introduced in the service domain 'navigation' in Generation 2 (see p. 78) of previous Chap. 2. However, in the future this will also include any event of interest like, a pothole, accident, weather situation etc. as enhanced in the service domain 'navigation' in Generation 3 (see p. 111). Today, we have companies providing this information to vehicles. In most vehicles, the information is still static. It will only increase in both frequency and the types of data in the future. Most data that are part of connected ADAS and V2V systems, like a car around the corner, end of traffic jams etc., will be crowd-sourced and shared with vehicles as well. There is lots of value in safety, but the business model for transmitting and sharing all this information has yet to be proven. However, an announcement[13] by HERE clearly show this direction and other OEMs will follow. Apple CarPlay and Android Auto could also be a threat in this area, certainly in low-end vehicles with basic IVI capabilities, but as the safety features and the HMI experience in the vehicle change, it will be much harder for Apple and Android to compete.

3.5.2   Internal OEM Business Domain Changes

The concept of owning the data and the value of the data is becoming very popular across all industries, not just automotive. IBMs CEO Virginia Rometty even back in 2011[14] talked about the value data and the insights that can be leveraged from it. Part of the automotive industry's transformation in the future seems to be to bring more of that data and the control of that data back in-house. Toyota,[15] Nissan,[16] and

---

[12]"Google's New Mobile App Cuts GPS Nav Companies At The Knees" http://techcrunch.com/2009/10/28/googles-new-mobile-app-cuts-gps-nav-companies-at-the-knees/. Accessed: July 26, 2017.

[13]"Mapping company Here will let drivers use maps from competing automakers" https://www.cnet.com/roadshow/news/mapping-company-here-will-let-drivers-use-maps-from-competing-automakers/. Accessed: July 26, 2017.

[14]"IBM's new CEO sees gold in data mining" https://www.marketplace.org/2011/12/30/tech/ibms-new-ceo-sees-gold-data-mining. Accessed: July 26, 2017.

[15]"Toyota, Microsoft Partner to Develop Car Tech That Drivers Want" http://fortune.com/2016/04/04/toyota-microsoft-data/. Accessed: July 26, 2017.

[16]"Renault-Nissan and Microsoft collaborate to deliver the future of connected driving" https://news.microsoft.com/2016/09/26/renault-nissan-and-microsoft-partner-to-deliver-the-future-of-connected-driving/. Accessed: July 26, 2017.

BMW[17] with Microsoft have had major announcements and investments in these areas. Sure, these types of systems are in the cloud and are announced together with IT providers, but it's the OEM that's making the investment and owning the data. In the future, it will be interesting to see how the smaller OEMs choose to invest in this field without having the funds typically available to the larger OEMs. Maybe as the traditional TSPs transition into supporting just Safety & Security services, they can look at ways of being more of a services aggregator and also providing the data to OEMs in a way where they have more access and control of the data. Certainly, technologies like the cloud and microservices can make this easier and offer a business model for the smaller OEMs to participate without making major investments.

The Generation 3 (2022) use cases and changing values of the OEM business domains were related to the cognitive edge and autonomous driving, see Fig. 3.16 for a summary followed by a detailed breakdown for each department.

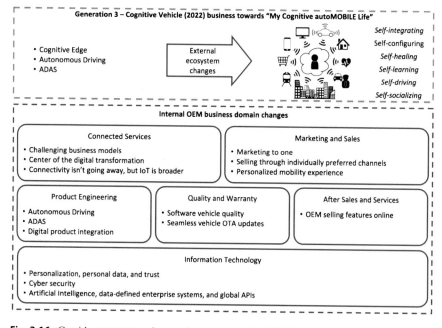

Fig. 3.16 Outside ecosystem changes impacting inside OEM business domain changes with Generation 3—cognitive vehicle (2022)

---

[17]"BMW partners with Microsoft to build a new kind of car intelligence" https://enterprise. microsoft.com/en-us/industries/discrete-manufacturing/bmw-partners-with-microsoft-to-build-a-new-kind-of-car-intelligence/. Accessed: July 26, 2017.

Connected Services

How successful will the new business models be? How will personalized vehicles be perceived? And will it be easier with more data to measure all those soft ROI factors?

With a new set of use cases come new challenges. Will safety and ADAS pay for wireless charges and allow for other scenarios to piggyback on cheaper wireless charges? Will merchant ecosystems and new business models pay so safety can be free? Either way, investments will be made related to managing the data and the connected service organization will be at the center of the digital transformation even as new mobility models come into play as well. The organizational challenges to leverage the competencies of Product Engineering and IT to continuously transform and scale the connected services from their own organization to the new ecosystem around the autoMOBILE (see Fig. 3.14).

Marketing and Sales

With the cognitive edge and a more personalized experience, marketing may just arrive at individual targeted campaigns with very personalized and targeted advertisements. As the OEM builds around the brand and a personalized experience, the amount of information and understanding of the customer increases dramatically, and marketing will need to shift to selling to individuals via their preferred channels throughout their life, whether a vehicle is involved or not. Hence, the personalized mobility experience will be the primary focus of marketing and sales.

Product Engineering

Product engineering is now on a journey to automate and personalize the vehicle. Most of the investments in autonomous driving and ADAS safety systems will go back to the traditional model, where the cost is covered in the price of the vehicle. But when certain safety features are driven by external data from other vehicles, traffic and weather sources, the business model will need to change. We mentioned HERE earlier, but how are the systems going to be paid for, specifically the wireless charges? These features will again put a lot of pressure on the connected services organization to be creative in covering the costs, maybe merchant interaction will be enough to cover the charges. Let us wait and see.

Now to the personalization aspect, not so dependent on large amounts of data over wireless, but from a trust and security standpoint, not to be taken lightly. The cognitive edge will drive more personalized features in vehicles, engineering will just need to work more closely with IT on how to manage this information in the cloud and internally with their CRM systems. The right business decisions towards digital product integrations to enable more adaptation of the vehicle after the sale will be most critical for the future and can make or break an OEM.

Quality and Warranty

Will OTA become as seamless as it is on the smartphone today? Let us wait and see, but the vehicle of the future will be as much about managing software as it will be about hardware. At least software and quality management are well-known processes and automotive can learn a lot from other industries, but the transition is sure to be difficult. The feedback from the machine learning sensors and the systems for ADAS alone will require continuous engineering and updating. But the savings and ability to manage without trips to the dealership will pay off in the long run.

After-Sales and Service

How will we engage in the future? Will we be more likely to buy new features for the vehicle online? Or will going to the dealership really survive? After-sales will certainly be change as part of the digital transformation. The physical maintenance does not go away, but will just be less of a focus, and will not be a differentiator in the market.

Information Technology

The center of IT investments for automotive in the future will be AI technologies, data-defined enterprise systems and global APIs. These technology investments are at the core of the digital transformation of all systems and future business interfaces. Cyber security principles are elementary for every designed digital service; physical protection like thick, high walls or locked doors do no longer exist. Protection happens at the software level only. IT needs to take more responsibility in transforming the traditional product engineering organization. For example, helping engineering with the development of the cognitive edge and managing personal data will be a focal area, along with leveraging vehicle and personal data for both internal and external users.

3.5.3   How Changes Result in Value?

Business models always seem to be a challenge, but internally everyone better understands the value of data and how to use them. Returning to a must-make-revenue strategy will not be successful. But looking at the customers and the revenue from them over a lifetime will be key, as personalization and brand become driving factors. The tools and technologies will actually help make this transition easier and more successful. But returning to the "creepy line" (see Fig. 1.5) we introduced in Sect. 1.1.2, how close will traditional OEMs come to endangering the established enterprise reputation and brand values (see Fig. 3.17)? Will new entrants into the market with much more IT expertise push traditional OEMs aside? The foundation for vehicle and consumer data insights is in place, the challenge remains which OEMs will approach it and how fast.

Conclusions

Personalization will help continue the shift in organizations and to move closer to the customer and the buying point. The in-vehicle ADAS will dominate the market, external digital data will be an option, but not to any real scale by 2022. However, as we have seen with transformations in IT and consulting over the last 10 years, big companies can move, albeit slow sometimes, but it can happen. Automotive

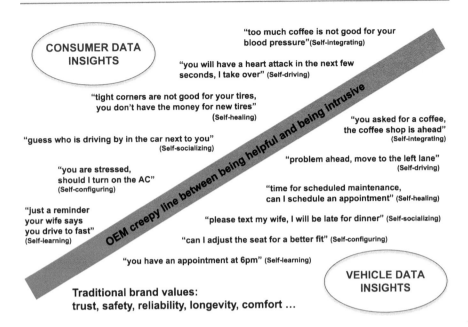

Fig. 3.17 The "creepy line" illustrates the breaking point between the two different business models where the enterprise reputation and brand values are endangering if the line is crossed

OEMs will transform the landscape and it certainly won't look like it does today. Remember it was only 10 years ago that the iPhone was introduced and look at the market and transformation in companies over that period.

Self-Enabling Values for Customer and OEM

We will review the self-enabling values for customer and OEM with Generation 3 (2022), (see Fig. 3.18), our first glimpse of a vehicle reaching towards its value goals. Remember our scoring was based on generic values, but as OEMs look into the future and design for the future based on the businesses and customers they want to support, they will need to set their scores accordingly. For example, our value for the customer around self-driving is medium. If we were a carsharing customer and wanted the additional safety and insurance cost reductions, the factor might be high. Similar for an OEM selling to fleet or commercial businesses that would factor this high for the value in increasing revenue and profit. But back to Generation 3 (2022). Customers now see the value in self-learning and a personalized experience with advanced CPAs and self-socializing vehicles making their daily commute more enjoyable. OEMs continue to see the value from real-time vehicle data and continue to grow new revenue sources.

Fig. 3.18 Self-enabling values for customer and OEM with Generation 3—cognitive vehicle (2022)

3.6   Generation 4: Cognitive Life

'My Cognitive autoMOBILE Life', where the vehicle is an integral part of a customer's personal network. That network includes the home ecosystem led by Amazon in the US, Allibaba in China, the smartphone ecosystem led by Apple, ridesharing led by Uber, mobility ecosystem led by Google, social media led by Facebook, and retail led by Amazon, just to name a few examples. Now what can automotive OEMs learn from all these different business models? We do not really care who is leading now or whether they will be leading in the future. Rather, we care about what business model got them there and what if anything automotive OEMs can learn from that. Figure 3.19 shows these top brands, their business model, and their core assets.

What is interesting is that all the companies that have strong assets related to software and cloud also have a strong focus on customer experience and

	Amazon	Google	Apple	Samsung
Business Model	Online Retail	Advertising	Product and Services	Products and Services
Core Assets	Cloud/Store	Cloud/Search	Cloud/iTunes, Devices	Cloud, Devices

	Alibaba	Uber	Facebook	Microsoft
Business Model	Online Retail	Ride Sharing	Advertising	Products and Services
Core Assets	Cloud/Store	Cloud/Delivery	Cloud/Networking	Cloud, Office

Fig. 3.19 Business models and core assets of leaders with ecosystems related to our life

personalization, and are very close to the customer buying point. Sound familiar? Imagine how much personal information each of these companies knows about their customers and now compare that to an automotive company.

Another interesting insight into these companies comes when from looking at the strategic markets around 'My Cognitive autoMOBILE Life'. Going back to Fig. 2.31, we see that three of these companies are looking at strategic moves into automotive, Samsung, Apple and Google. Six of the companies have autoMOBILE capabilities, and nine of the companies are developing CPAs. Is this a race to see who can fill in all the boxes first? Will anyone new come along? Will we trust these brands near the creepy line?

An important part of rethinking the entire business model for the cognitive life is the customer's e-motion through life. With a series of experiences before the customer even realizes the need, the autoMOBILE companies are there through the CPA to provide the trusted links to ecosystems and suitable advice for the experience ahead. In Fig. 3.20, we have illustrated that entire business model related to our cognitive life.

We will start with the customer at the center, buying services and products from all types of providers. The first example is the smartphone with CPA services along with the application ecosystem, actually in some cases another connection to other brands, but for simplicity, we will not include that. The customer can also buy home appliances, home automation products and services from OEMs or new cognitive-life brands such as Amazon. Vehicles and connected services can be purchased from dealers/OEMs. Hotels, airlines, city buses and city train services are purchased as customers travel locally or regionally. New mobility providers will provide services and online retailers will provider all types of products for customers to purchase. In most of the cases, we included a dashed line where the customer can interact with and also purchase using the CPA on the smartphone. This is not really associated with the business model other than to show the focus and control of the CPA, especially as brands start to change and move into other domains. The other business models we previously discussed still apply as well. The real effect in the cognitive life is when a brand enters a domain and basically pushes an existing brand out. Can anyone name the bookstore chains in the US from 10 years ago?

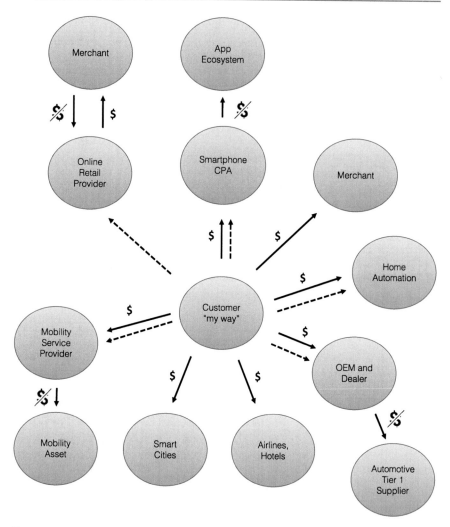

Fig. 3.20 Flow of money and interactions (dashed line) in an entire business model linking ecosystems related to our cognitive life

3.6.1   External Ecosystem Changes

▶   Can automotive refocus on services and building a brand?

We are starting to see some examples. Ford has taken it one step further with the Ford mobility experiments making core data available to an ecosystem of application developers.[18] Look at some of the ideas and concepts, painless parking, dynamic social shuttle, data-driven insurance, Ford carsharing, car swap, parking spotter, etc. Does this sound like a traditional OEM? This strategy aligns with their investment in SmartDeviceLink, also making vehicle controls available to mobile application in the vehicle.[19]

Ford sees the value in being an enabler, others may see that value as a new route to market. Take Uber for example. If more vehicles were enabled with the capabilities like Ford, their application could be more integrated into the vehicle and ultimately provide better services to their customer, whereas today the data they use are only available from the phone. More and more, as we move towards autonomous, the need to have data from the vehicle will be important and Ford must see that and is experimenting to understand how they will enable this more in the future. Is their strategy better than others, is this the path to a cognitive vehicle and a personalized mobility experience? It won't seem so, since they are giving access to the data to others, but if we look at their FordPass[20] mobile application, we can start to see what a cognitive vehicle will be in the future, and the value in investing in value-added devices. To take a quote from their website,

At Ford, we aim to keep you moving freely. The FordPass App makes your on-the-go easier. Our new FordHub reimagines the future of transportation. And our friendly FordGuides are there to help every step of the way. Join us, and discover what we mean when we say Go Further.

Does this sound like a car company? What happened to engine horsepower? FordPass Perks, clearly a move towards brand and loyalty over traditional vehicle style. The FordPass app, with parking, vehicle controls, digital wallet, credit, the user can really start to see the potential of the move to a cognitive life now.

Remember in [26] how OEMs change is key to offer these mobility services, but is Ford on the right path? Organizationally, they are moving to the autoMOBILE value chain (see Fig. 1.3), sales, digital access and value-added services can be seen in their strategy. An autonomous "minimalistic travel capsule" can really only be differentiated by some other set of services or capabilities, are Ford and others starting down this path already? Only time will tell if this is the right strategy.

Uber certainly has shown the value of ridesharing, and most of the aforementioned ideas are new services and sources of revenue for OEMs, so they are moving towards more value and connection to their brand. Customers will see the value as

---

[18]"Ford at CES Announces Smart Mobility Plan and 25 Global Experiments Designed to Change the Way the World Moves" https://media.ford.com/content/fordmedia/fna/us/en/news/2015/01/06/ford-at-ces-announces-smart-mobility-plan.html/. Accessed: July 26, 2017.

[19]"New SmartDeviceLink Developer Program Makes It Easier to Create In-Car Apps for Growing List of Global Automakers" https://media.ford.com/content/fordmedia/fna/us/en/news/2016/06/10/ford-sdl.html/. Accessed: July 26, 2017.

[20]"FordPass – A Smarter Way To Move" http://www.fordpass.com/. Accessed: July 26, 2017.

well, with more options for mobility in the future, the potential in cities to reduce the monthly cost of transportation, whether the consumer owns a vehicle or not.

Mercedes, a bit different than Ford, is investing in service enablement with their recent redesign and launch of Mercedes me.[21] This isn't just a new website, it's the initial step to building a brand and connecting that brand to the customer, and what better word to do that than 'me.' The words we read when we first get to the site say it all,

> your personal point of entry into the world of Mercedes-Benz. Discover innovative services and fascinating offers,

and services like connect, assist, finance, inspire and move—there is no link to actually buying a vehicle—it's all about a personalized customer "my way" experience. Let us look at a few of these services in more detail.

Mercedes me connect:	This goes beyond the typical Safety & Security, it follows the concept to be always connected to the owner's vehicle, a much different approach.
Mercedes me move:	Something similar to Ford, but the services are promoted through different brands. They include: car2go, one of the first globally available carsharing services, moovel, a mobility planning smartphone application that offers options for taxi, public transit, etc. and the ability to buy one ticket for multimodal travel, making owning a car unnecessary, Blacklane, a more exclusive Uber, and MyTaxi, a clear competitor to Uber.
Mercedes me assist:	A new look at service and what the customer needs to do to keep the vehicle in peak performance. A twist on maintenance as well, where the customer's digital service report is securely kept for the service life of the owner's vehicle, no matter what dealership is used.
Mercedes me finance:	It's hard to put any new twists on the basics on financing and leasing, but Mercedes also offers insurance as well, but no new types of models such as PAYD.
Mercedes me inspire:	A lifestyle configurator, what else can we say? Is this an automotive website? Finally, the place to find out about vehicles, but it's much more than that. It is events, promotions, motor racing, their vision of electric vehicles, and even a championship golfer returning a trophy in a Mercedes concept vehicle.

This new type of branding also flows over into their new storefronts such as Mercedes Connections in Tokyo. Your first impression is that it is a coffee shop or a

---

[21]"Mercedes me – Mercedes-Benz" http://mercedes.me. Accessed: July 26, 2017.

restaurant, but it's really an OEM-sponsored store with vehicle displays, test drives, merchandise and even loyalty points. It is nothing like their traditional dealerships in the US. We have even seen special events like the Oktoberfest and activities in the parking lot, now this is e-motion and a new way to connect with customers.

3.6.2   Internal OEM Business Domain Changes

The Generation 4 use cases and changing values of the OEM business domains were related to personal networks and mobility. See Fig. 3.21 for a summary followed by a detailed breakdown for each department.

Connected Services
It's all about digital content and context, a personalized experience and mobility applications using global APIs to integrate the personal life. The merchant ecosystem and other brand affiliations will be behind most of the CPA is leveraging. Possibly even cities and public transportation will become more integrated to provide a true mobility experience. What was the connected vehicle is now a cognitive life, people connected in ways we don't even see yet. The automotive industry will be an important part of that ecosystem, but not the center of the life.

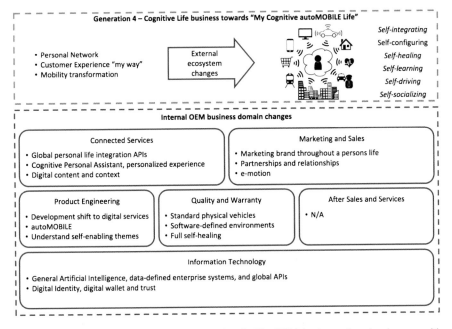

Fig. 3.21 Outside ecosystem changes impacting inside OEM business domain changes with Generation 4—cognitive life

Marketing and Sales

The transformation of marketing to targeting individual customers throughout their lives and probably their type of lifestyle will be the focus. Marketing will look into the partnership and relationships each OEM will want to have with other brands, the brands that are associated with the types of lifestyle they are looking for, aligned to their own brands. With the e-motion, will we stay aligned with automotive brands in the future, like we align with smartphone brands today?

Product Engineering

Now that vehicles are finally fully autonomous, they really do not look very different anymore, just like in the aviation industry, where strict regulations have resulted in the cockpit of one aircraft resembling that of all others. Most of the development and research focus will clearly shift to digital and mobility applications. The organization acquired the skills to develop the cognitive autoMOBILE by considering all six self-enabling themes. The vehicle interior is in designed such a way that it can change dynamically on customer demand without buying a new product.

Quality (and Warranty)

With the standardization on the physical side due to autonomous driving, the focus on quality and distribution in the software-defined environment will be the top priority. All OTA update and self-healing capabilities are provided by global APIs. Warranty offerings exist in previous generations because product engineering is concerned that it has not done a good job. That is now changing with self-healing and the organization can be renamed to just 'Quality.'

After-Sales and Service

Merged into the connected services team as fully automated services accessed through global APIs.

Information Technology

Only the OEMs who intelligently organized their IT, product engineering, and connected services will make it to this Generation 4. The entire business is defined by APIs, public or protected, internal or external, global or regional, data or transaction-oriented, vehicle or mobility-related, providing business capabilities or technology enablers. Strategic investments in new technologies like a distributed ledger are becoming relevant to establish transparency and trust for the customers sharing personal information through the digital services. IT will be mission critical for the new business around 'My Cognitive autoMOBILE Life'.

3.6.3 How Changes Result in Value?

Cognitive-life brands that want to grow will need to continue to gain more control of customer touchpoints. Interesting are some of the strategic business moves each company is making. Amazon is creating more channels to its new version of retail.

Echo isn't perfect yet, it's just a channel to the core business. Google is creating more channels to search, the Android OS, Google Home, and their applications, all are coming back to the search feature. Apple has more devices, all coming back to content and service on iTunes. Samsung has more devices, all with the potential integration with their smartphones. Uber has started new businesses like UberEATS and Uber Freight that utilize their core software capabilities in the world of mobile and on-demand delivery. So what are the new channels leading back to the automotive industries core assets? Certainly ridesharing and carsharing, followed by a very personalized mobility experience, and connections to other forms of mobility and the travel industry. All these companies have a highly recognizable brand and their products have not become commonplace, so as automotive transforms, it will need to keep its products differentiated. Can the automotive companies compete if Apple and Google enter the market? An automotive company can only dream about the profits Apple makes on just the hardware. That is what's scary when we think about an Apple car. It won't be hardware with a small profit margin. The reason for this is that Apple will also implement a digital platform around the vehicle, just as it did with the smartphone. This will allow third-party developers to build applications around their vehicle platform and probably match the success of their smartphone business. Are OEMs ready to do this? Let us see what happens with BMW, Ford, and General Motors in the near future.

Conclusions

Automotive OEMs are in an interesting position as their industry transforms around mobility, automation and a cognitive life. OEMs that can grow their brand, grow their touchpoints with customers and grow in the new mobility industry will remain strong in the automotive industry. For cognitive life, they will need to collaborate correctly with other brands to better position themselves to compete or join with cognitive-life brands in the new personal network.

▶ Who Says Elephants Can't Dance?

is the name of a book [1] by former IBM CEO Louis V. Gerstner, who took over in 1993 when many felt IBM was no longer a viable player in the revolutionized computer industry. However, he turned around the company. Now can elephants dance in the automotive industry revolution? For the automotive OEMs that are too slow to change, they will eventually disappear or become a tier 1 supplier to cognitive-life brands that need the travel capsule of the future.

Self-Enabling Values for Customer and OEM

We will review the self-enabling values for customer and OEM with Generation 4 (see Fig. 3.22). The customer experiences the value of a vehicle that is now more like a smartphone and self-integrating and self-configuring are seamless around their lives. The value for self-driving is going beyond safety and becoming a major convenience and enjoyable on both short trips with traffic and long trips on open

Fig. 3.22 Self-enabling values for customer and OEM with Generation 4—cognitive life (2025)

roads. OEMs have realized the value of sharing vehicle data and personal data to improve the experience in the vehicle and increase revenue.

Our first glimpse of a vehicle approaching its value goals. Remember, our scores were based on generic factors, but as OEMs look to the future and design for the future based on the businesses and customers they want to support, they will need to set their scores accordingly. For example, our value for the customer related to self-driving is medium. If a carsharing customer wants additional safety and insurance cost reductions, the value might be high. Similar for an OEM selling to fleet or commercial businesses that would score this as high for the value in increased revenue and profit.

Customers now see the value in self-learning and a personalized experience with advanced CPAs and self-socializing vehicles making their daily commute more enjoyable. OEMs continue to see the value from real-time vehicle data and continue to grow new revenue sources.

References

1. Gerstner LV Jr (2002) Who says elephants can't dance?: leading a great enterprise through dramatic change. Harper Collins, New York, NY
2. Kuhrmann M, Münch J et al (eds) (2016) Managing software process evolution: traditional, agile and beyond – how to handle process change. Springer, Cham
3. Leach RJ (2016) Introduction to software engineering, 2nd edn. CRC Press, Boca Raton, FL
4. MBtech Consulting (2011) Trend analysis: connected car 2015 – the most important trends and challenges in vehicle telematics, Sindelfingen. https://www.mbtech-group.com/fileadmin/media/pdf/consulting/downloads/Trendanalyse_Vernetztes_Fahrzeug_2015_EN.pdf. Accessed 26 July 2017
5. Mingay S, Mesaglio M (2016) Deliver on the promise of bimodal. Gartner, Stamford, CT
6. Perkin N, Abraham P (2017) Building the agile business through digital transformation. Kogan Page, London
7. Wedeniwski S (2015) The mobility revolution in the automotive industry: how not to miss the digital turnpike. Springer, Heidelberg

Challenges Ahead in 'My Cognitive autoMOBILE Life'

4

The challenges ahead in 'My Cognitive autoMOBILE Life' starts with a story. The context is related to

1. the cognitive vehicle and the industry transformation as we introduced in Chap. 1,
2. the technologies and disruptive innovations in the automotive industry we analyzed in Chap. 2, and
3. the evolving ecosystems around our life and the values we discussed in Chap. 3.

With this context, our story is illustrated in Fig. 4.1. As through all three previous chapters, this story is about the foundational six self-enabling themes (see Sect. 1.6)

- Self-integrating,
- Self-configuring,
- Self-learning,
- Self-healing,
- Self-driving,
- Self-socializing

to look at the challenges ahead as we move towards a cognitive life.

Let us call him Henry. He enjoys his hectic work environment, but also likes to get outdoors and enjoy the adventure of nature. He has grown up in a technology-friendly world and embraces technology as long as it is secure and his privacy is protected. Henry is married and has two daughters. The family loves time spending together at places where dreams come true.

The journey starts as Henry is about to buy a new vehicle for business and personal use.

© Springer-Verlag GmbH Germany 2017
S. Wedeniwski, S. Perun, *My Cognitive autoMOBILE Life*,
https://doi.org/10.1007/978-3-662-54677-2_4

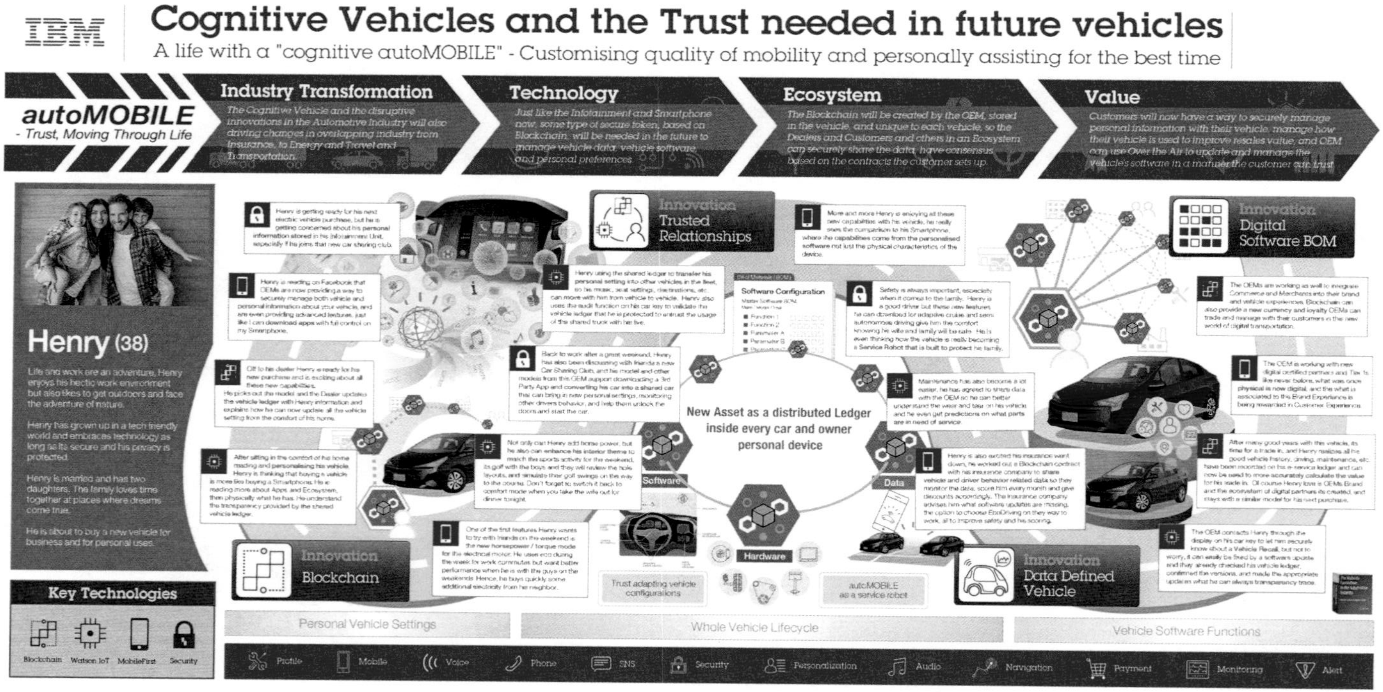

Fig. 4.1 Customer journey map in the age of blockchain to establish trust and integrity in future vehicles (Source: IBM)

1. Self-integrating: Henry is getting ready for his next electric vehicle purchase, but he is getting concerned about his personal contacts stored in his current head unit, especially if he joins that new carsharing club.

2. Self-integrating: Henry is reading on Facebook that OEMs are now providing a way to securely manage both vehicle and personal information about your vehicle, and are even providing advanced features, just like I can download apps with full control on my smartphone.

3. Self-integrating: Off to his dealer, Henry is ready for his new purchase and is excited about all these new capabilities. He picks out the model and the dealer updates the vehicle ledger with Henry's information and explains how he can now update all the vehicle settings from the comfort of his home.

4. Self-configuring: After sitting in the comfort of his home reading and personalizing his vehicle, Henry is thinking that buying a vehicle is more like buying a smartphone. He is reading more about apps and ecosystems than what he owns physically. He understands the transparency provided by the shared vehicle ledger.

5. Self-configuring: One of the first features Henry wants to try with friends on the weekend is the new horsepower / torque mode for the electric motor. He uses eco during the week for commuting to work, but wants better performance when he is with the guys on the weekends. Hence, he quickly buys some additional electricity from his neighbor.

6. Self-configuring: Not only can Henry add horsepower, but he also can enhance his interior theme to match the sports activity for the weekend. It is golf with the boys and they will review the hole layouts, and simulate their golf swings on the way to the course. Don't forget to switch it back to comfort mode when you take the wife out for dinner tonight.

7. Self-configuring: Back to work after a great weekend. Henry has also been discussing a new carsharing club with friends, and his model and other models from this OEM support downloading a third-party app and converting his car into a shared car that can incorporate new personal settings, monitor other drivers behavior, help them unlock the doors and start the car.

8. Self-configuring: Henry uses the shared ledger to transfer his personal settings to other vehicles in the fleet, so his music, seat settings, destinations, etc. can move with him from vehicle to vehicle. Henry also uses the audit function on his car key to check the vehicle ledger and confirm that he is safe when he entrusts the shared truck with his life by using it.

9. Self-learning: More and more, Henry is enjoying all these new capabilities of his vehicle. He really sees the comparison to his smartphone, where the capabilities come from the personalized software, not just from the physical characteristics of the device.

10. Self-driving: Safety is always important, especially when it comes to the family. Henry is a good driver, but these new features he can download for adaptive cruise and semi-autonomous driving give him peace of mind knowing that his wife and family will be safe. He is even thinking about how the vehicle is really becoming a service robot that is built to protect his family.

11. Self-healing: Maintenance has also become a lot easier. He has agreed to share data with the OEM so he can better understand the wear and tear on his vehicle and he even gets predictions on what parts are in need of service.

12. Self-socializing: Henry is also excited his insurance went down. He worked out a blockchain contract with his insurance company to share data related to vehicle and driver behavior so they can monitor the data, score him every month and grant discounts accordingly. The insurance company advises him what software updates are missing, the option to choose EcoDriving on the way to work, all to improve safety and his score.

13. Self-healing: The OEM contacts Henry through the display on his car key to let him know securely about a vehicle recall. But not to worry, it can easily be fixed by a software update and they already checked his vehicle ledger, confirmed the versions, and made the appropriate updates, which he can always transparently track.

14. Self-socializing: After many good years with this vehicle, it is time for a trade-in, and Henry realizes all his good vehicle history, driving, maintenance, etc. have been recorded on his e-service ledger and can now be used to more accurately calculate the value for his trade-in. Of course, Henry loves his OEM's brand and the ecosystem of digital partners it has created, and stays with a similar model for his next purchase.

15. Self-socializing: The OEM is working with new digital certified partners and tier 1 suppliers like never before. What was once physical is now digital, and what is associated with the brand experience is being rewarded in customer experience with increased vehicle usage.

16. 'My Cognitive autoMOBILE Life': The OEMs are also working on integrating commerce and merchants into their brand and vehicle experiences. The blockchain can also provide a new currency and loyalty, which OEMs can trade and manage with their customers in this new world of digital transportation.

This journey leads us into the discussion of the level 4 and level 5 issues for each of the six self-enabling themes and what some of the current market players are doing. We will also touch on some technology options that OEMs and tier 1 suppliers can investigate. We will finish with the self-socializing level 5 *e-motion*, and the trade-offs cognitive life brands will have to make moving forward.

4.1   Blockchain

In 'My Cognitive autoMOBILE Life', when we look at the technology enablers and business models we discussed in the previous chapters, trust and privacy is essential through all six self-enabling themes. One technical capability that solves many of the challenges the industry will face in the future related to trust and transparency of this information is blockchain technology. When we first started talking to customers about this idea, we created the journey map (see Fig. 4.1) to walk through a few scenarios to explain, from the customers point of view, where and when blockchains can be the foundation for the self-enabling themes related to the cognitive vehicle.

> A *blockchain* is a distributed database that maintains a continuously growing list of ordered records called blocks. Each block contains a timestamp and a link to a previous block. By design, blockchains are inherently resistant to modification of the data. The data in a block cannot be altered retroactively once recorded.

The advantages of blockchains are seen as an open, distributed ledger that can record transactions between two parties efficiently and in a verifiable and permanent way. For a more detailed introduction and explanation of blockchains, we refer to [3] and [8], for example.

Just some basics of blockchains that can be used as core capabilities in the cognitive vehicle. First word that comes up is distributed database. It has been around for a long time, nothing really new, but the core capability needed to share information. The next capability, resistant to modification once recorded. Now we have a history or a record of everything that was done to the shared information. We put the history together in a chain with the information being in a block, hence the term blockchain. Still nothing really new, so the next capability is what is called a distributed ledger that is simply a record of transactions. An example of a simple transaction is a business selling a car to another business. There would be a transaction that moved the asset from one business ledger to another business's ledger. If the transaction had conditions, for example, there was a contract that stated the car must start for the transaction to be complete, then we have transactions with conditions, which in blockchains are called smart contracts. The information that is being shared are transactions against a distributed ledger including smart contracts, and are resistant to modification once recorded. The last attribute is consensus, which is what is needed when a change is made and the others in the network agree to the change as well. In the case of a distributed network, different nodes have this capability versus a single node in a centralized system. Finally, we need a secure environment because of the described contracts and transactions. Simplified, user authentication is used for all participants in the network and data in the blockchain can be visible to everyone or only based on access control. Data within a blockchain can be encrypted. We already mentioned data integrity and how the blockchain history cannot be modified.

4.2 Self-Integrating

In 'My Cognitive autoMOBILE Life', many scenarios talk about personalization, payment, identification and personal information. There are many approaches and methods for addressing trust and privacy. Every company takes their own approach with a combination of terms and conditions, privacy policies and IT implementations. In the past, ownership and understanding of vehicle data has led great confusion and discussion among vehicle owners and OEMs. As the vehicle moves through its life cycle from manufacturers and dealers to multiple owners, who owns this data and when it can change? A personal vehicle ledger, unique and stored in each vehicle, can be the new ownership, privacy, and control for this information. A blockchain ecosystem would share business transactions against assets changing the vehicle configuration that is stored in a distributed vehicle ledger for each vehicle. The personal vehicle ledger is the new digital asset that can contain information about a vehicle including the owner, the owner's personal settings, preferences, real-time vehicle data, and the vehicle's software BOM. This could be the foundation leading to a software-defined vehicle platform (see Fig. 1.4).

Blockchain technology is just one approach that could solve this problem, but the underlying issue of trust and transparency will be an important factor in the future, not just in automotive, but in the cognitive life. For PAYD insurance, once the insurance company has our data, who's not to say they won't use it in a way we didn't expect. We have control of giving it to them, we trust them, but that doesn't stop all the problems. It seems people are currently okay with what Google and Facebook are doing with personal data, but as more and more companies approach the creepy line (see Fig. 3.17), the number of companies that really can be trusted is likely to become smaller. Issues of data breaches are already making it into the press, and in the future something that can destroy a brand reputation very easily.

In this new world of data driven systems, privacy will be a big concern and the customer should not lose control over it. Google, Facebook and smartphones have redefined privacy and how personal data are used, not always with trust and transparency, bringing us back to the creepy line. Who reads those terms and conditions anymore anyway? But as we move forward towards a cognitive life, let's look at the two main data-driven solutions around the automobile, where the vehicle data and the personal data about us in the vehicle can and will be used in the future. We own that data, at least this is our perception, so if we give it up, we want to make sure we are getting something in return. In this context, there are many questions, which are related to the classification of data, claims to this data and liability for incorrect processing of data. Some examples include:

- Which data belongs to the driver because (s)he is the one using the vehicle?
- Which data belongs to the passengers because they are traveling in the vehicle?
- Which data belongs to the owner or keeper of the vehicle?
- Which data belongs to the manufacturer or to the supplier, because they are both liable for the vehicle which has been constructed?
- Which data belongs to the public or to the state if the vehicle is located on a public road, or is involved in a crime?
- And how does all of this change when the car is taken to another country?

More detailed information on data protection law discussions related to the connected vehicle is available in the literature [1, 6].

Don't forget that beyond sharing personal information, vehicles are also sharing information, even in real time. Some of these systems already exist today. We talked about the majority of them and the companies that are leaders in the market today. Google, Waze, TomTom, and Inrix all have solutions in the market today to collect and distribute crowd-sourced data for traffic and incidents while we are driving. The time it takes to get updated information to us is typically 5–10 s and works in most cases. But we will discuss further how this is changing in the future.

An example of the next step will be Waze and SmartDeviceLink, where along with traffic information, the vehicle data will be included as well.[1] This approach enables new applications to get access lots of vehicle data quickly, due to the number of existing vehicles with the SmartDeviceLink solution. Beyond Toyota and Ford, Daihatsu, Mazda, PSA, Subaru and Suzuki have signed up and will launch vehicles with SmartDeviceLink as well. Android Auto can also support similar features. Apple CarPlay will only support Apple Maps at this time, which is again a large number of existing vehicles in the market today. The final step in vehicle sharing and also supporting more autonomous driving scenarios would be the HERE solution we discussed earlier (see p. 111).

4.3 Self-Configuring

How custom can we make a vehicle, will it need to be custom in the future of travel capsules? These decisions first need to be separated into two categories, mission critical and the digital platform (see Fig. 2.26). Mission critical are the systems in the vehicle that are 100% required to provide a safe driving experience. The digital platform is what we personally bring into the vehicle and part of our personalized experience. The challenge comes when the digital affects the mission critical, when we want to change the configuration of our vehicle from a personal vehicle to a ridesharing vehicle, it still must be safe to drive.

Before we talk about mission critical, let us look at the phases of how some of these new types of configuration will enter the industry. We can already see today some simple configuration changes that can be made just by identifying ourselves to the vehicle, either as driver 1 or as an individual person. The next phase we introduced is the digital cockpit (see Fig. 2.24), a great example of the newer types of digital changes that can be brought into the vehicle. We pushed the industry with a level 5 goal for 3D printing of parts or components at the dealer. This might be a stretch, but the fundamental point even in the short term is how do we deal with offering configurations even to the point where parts are printed at the dealer. This isn't that far from some of the dealer-installed options today, but automotive and other industries are already using this technology where it can be cost effective,

---

[1]"SDL Puts Waze Navigation, ParkU and Honeywell Home Thermostat Control Right on Your Dashboard" https://smartdevicelink.com/news/sdl-puts-waze-navigation-parku-and-honeywell-home-thermostat-control-right-on-your-dashboard/. Accessed: July 26, 2017.

see [2] for more details in these industries. Even in the medium term and in some regions around the globe, the physical aspect of the vehicle isn't going away, so to be able to adopt and offer these types of configurations can be valuable in the future.

Now back to the example of changing the vehicle configuration from a personal vehicle to a ride sharing vehicle. At first glance we really don't see any issues, but when we get to the level of, did the critical systems change in any way, this is when we have to look at the problem a bit differently. Take Google for example, the architecture on the smartphone from the device drivers, to the operating system and up to the applications, this is the fundamental architecture design for computer for the last decades. Now compare that to real-time automotive system, not quite the same. This will certainly be one of the areas in which the cognitive life brands will need to develop skills beyond their current software capabilities. Maybe a reason why Samsung acquired Harman. Either way, the concepts of the vehicle as a digital platform do create boundaries that brands will need to separate clearly as they move forward with a cognitive vehicle and into a cognitive life.

Blockchain technology could be used to support both types of configuration changes in the vehicle. Let us take the ridesharing example first, where blockchains could manage the contract between Henry and his insurance company (see Fig. 4.1). Once the vehicle is switched into ridesharing mode, the insurance company can get the required data it needs to offer Henry a better policy while he is working. For the mission-critical software, the blockchain can include a software BOM with the version numbers to provide integrity for the vehicle's critical software. At the appropriate time, the vehicle can sync with the OEM and get the OTA updates, and with the transparency and ledger of the blockchain, Henry can view and see a history of the changes.

4.4   Self-Learning

With all the hype around CPAs and all the related announcements, there are still fundamental issues in automotive where voice recognition leaves a lot to be desired.[2] The challenge is to bring the NLP capabilities of a cloud connected CPA to a disconnected in-vehicle voice recognition system. Certain OEMs and low-tech cars will not address the problem and will use smartphone integration to solve the problem. It's simpler and cheaper to use NLP on the smartphone and the connection to the cloud. Higher-end vehicles will move to a hybrid approach, where the in-vehicle experience is improved by also being connected to the cloud. Even with hybrid, another challenge is to build the domain information around the vehicle itself. Try asking Siri to help us change the tire. What about other vehicle domains like safety, trust, transparency, multiple occupants, cognitive spaces, attention of

---

[2]"Bluetooth Connectivity, Voice Recognition among Top 10 Problems in 3-Year-Old Vehicles" http://www.jdpower.com/cars/articles/jd-power-studies/bluetooth-connectivity-voice-recognition-among-top-10-problems-3-year. Accessed: July 26, 2017.

the driver, mood of the driver, and connected brand lifestyle in an AUTOmobile context? The combustion-engine-oriented AUTOmobile is possibly too complex to build an ecosystem around (see Fig. 3.14) that will not lead the market towards 'My Cognitive autoMOBILE Life'. The cognitive vehicle architecture first needs to be changed (see Fig. 2.26), then the CPA will become an integrated part of the design instead of being an add-on to the AUTOmobile.

Let's challenge the industry today and see how major players are positioned in the field of cognitive vehicles and CPAs. Recently, Amazon and Ford made an announcement and demonstrated the Alexa integration into a mobile application running SmartDeviceLink and running in the vehicle on a Sync 3 IVI.[3] The scenarios demonstrated were not like the one above. They had the driver put Tide laundry detergent in their Amazon cart, they enabled users to listen to audiobooks inside the vehicle, search and transfer local destinations to navigation, request news, and play music. They also demonstrated using Alexa from the home to start the vehicle remotely, or say commands from the vehicle to integrate with the home to close the garage door. We really didn't see a lot of innovation, most of the functions are available today. Amazon is just adding a voice interface upstream of them. Lacking were the capabilities to have a conversation and learning, i.e. say more than one type of command or request, the capability to be event-driven and collect data for an analysis of the driving style. Not that Ford and Amazon won't do these things in the future, but the capabilities weren't demonstrated so far. We also wonder in this case who is making the investment and who is gaining the value, it seems Amazon is gaining most of the value and is probably making most of the investment. We see some of these scenarios being of value to consumers, but let the hype settle and see how the production version works and performs with regard to learning.

Similar to Amazon, Google together with Hyundai announced the capability to send commands from Google Home to the vehicle.[4] From an innovation standpoint, nothing really new, but we can assume other capabilities are in the planning stages as well, and with Android Auto and the Android operating system becoming very popular in the industry, tighter ties to Google Assistant from inside the car will certainly be seen in the future. Recently, Google and Chrysler also announced joint work on the Android operating system and integration with the Google Assistant in the vehicle.[5] The details of what was embedded or what was in the cloud are unknown so far, but clearly movement in the future.

---

[3]"Ford turns Amazon Alexa into a driver assistant" https://www.cnet.com/roadshow/news/ford-turns-amazon-alexa-into-a-driver-assistant-echo-ces-2017/. Accessed: July 26, 2017.

[4]"Hyundai links up with Google Assistant for car voice commands" https://techcrunch.com/2017/01/03/hyundai-links-up-with-google-assistant-for-car-voice-commands/. Accessed: July 26, 2017.

[5]"Fiat Chrysler and Google team up on a concept automotive infotainment system powered by Android" http://www.androidpolice.com/2017/01/02/fiat-chrysler-google-team-concept-automotive-infotainment-system-powered-android/. Accessed: July 26, 2017.

Microsoft also made announcements with BMW and Nissan to bring Cortana into the vehicle.[6] A small demonstration was shown in the BMW vehicle, but if we look at Microsoft's cloud and Cortana today, we easily see all the capabilities needed for a CPA. How their conversational and natural language capabilities stack up in the market will be left to experts in those fields, but we will wait to see what their production systems looks like. With Cortana's foundation on Windows, the ties to office productivity, mail, and documents, along with their acquisition of LinkedIn, their type of system will clearly have more of a business focus than the current shopping focus of Amazon.

We consider Samsung to be one of the top players in the industry based on their recent acquisitions of Harman and ViV. Combining this with their expertise in the smartphone and appliances industry, already working with Google and Android, they clearly have all the components in place.

Another challenge for the CPA providers is building an ecosystem and getting brands to participate. All five companies are building what they call skills or actions around their ecosystems, more than 15,000 on Amazon for example, that's a lot of brands for news, weather, sports, shopping, connected car, health and fitness, music and audio, social, etc.[7] It might be a challenge to remember all the skills of the CPA to understand what the user really wants. NLP helps, but once the user gets past news, weather and sports, the phrases or commands for certain skills would become pretty unique and he loses the point about having a natural conversation. An area that will certainly need to be improved and surely they are all working on it, but currently this looks to be more of a framework for advanced commands than a natural conversation system.

Channels or access to the customer will be a challenge as well. All the providers are making strategic moves in this direction. Special hardware or devices at the edge will be the battleground to lock the customers into the personalized channels to consume the services. Apple is already a clear leader in this field including smartphones, tablets, PCs, CarPlay, home devices, and Apple TV. Amazon wasn't really even in the CPA industry until they introduced Echo. Microsoft is building on its strengths in tablets and PCs. The big three in the smartphone market, Apple, Samsung, and Google will have a huge advantage, and maybe even Microsoft with Lumia.[8] How Amazon will get outside the home is open yet, but its rumored that the next Echo will have a screen. It is not a stretch, considering most people are already using tablets in the kitchen, why not make one for that purpose alone?

---

[6]"CES 2017: Microsoft's Cortana Finds New Homes In Nissan, BMW Cars Via Connected Vehicle Platform" http://www.techtimes.com/articles/191634/20170107/ces-2017-microsofts-cortana-finds-new-homes-in-nissan-bmw-cars-via-connected-vehicle-platform.htm. Accessed: July 26, 2017.

[7]"Alexa Skills Kit—Build for Voice with Amazon" http://developer.amazon.com/alexa-skills-kit. Accessed: July 26, 2017.

[8]"IDC: Smartphone Vendor Market Share, 2017 Q1" http://www.idc.com/promo/smartphone-market-share/vendor. Accessed: July 26, 2017.

The edge hardware is just at the beginning to digitize entire industries and especially the autoMOBILE. Today, the most important market of wearable devices is that of watches, especially the ones including third-party applications like Apple Watch or Samsung Gear. Example applications leveraging the digital watch in the connected vehicle field like unlocking the car or providing vehicle information already entered the market in 2015 from Mercedes and BMW. However, in the era of the cognitive vehicle and cognitive life, the range of new cognitive edge hardware in ear wear and clothing will grow significantly with many new use cases to fully integrate the CPA into our life.[9]

Which provider will be the first to work with an OEM and build the automotive domain knowledge? A domain knowledge, for example, is what IBM did with Watson for Oncology. It builds the big data store for all the information available about cancer and then provides a contextual dialog and machine learning around it to gain knowledge or provide recommendations.[10] For a more in-depth look at health knowledge, we refer to the literature [7]. This is not a simple skill or command like "order my pizza" or "what is the weather outside". It is a natural language conversation to get a recommendation for a problem, very different and much more complicated to build, even after billions in investments. Healthcare can be seen as a similar domain knowledge entry as automotive.[11] Nowadays, all solutions have the commands for "unlock door" or "start the engine", but this is not the vehicle domain knowledge where, for example, instructions on how to change the tire or to help the driver to pair his Bluetooth phone cannot be uttered in a single phrase. This is why even very strong brands such as Mercedes still work with Google and Amazon for the basic commands. However, they have to look into deep domain knowledge as well.[12]

As we look at CPAs and the future autoMOBILE, where is all this machine-learned personalization going to be stored and how will it be managed? As we expounded, the cognitive edge (see p. 113), blockchains could be used to store and manage the contextual personal information of all the different types of vehicles that we may use in the future. The concept of a smartphone application that is a client in the blockchain network that lets the user view and manage this information transparently can be very valuable. In addition, blockchains can help to keep this personal information secure, with the information owner in control of who accesses it, and distributed between all the blockchain nodes around the cognitive life.

---

[9]"Worldwide Wearables Market to Nearly Double by 2021, According to IDC" http://www.idc.com/getdoc.jsp?containerId=prUS42818517. Accessed: July 26, 2017.

[10]"IBM Watson Health—IBM Watson for Oncology" https://www.ibm.com/watson/health/oncology-and-genomics/oncology/. Accessed: July 26, 2017.

[11]"IBM Watson AI Suffers From Hype Despite Healthcare Potential" http://fortune.com/2017/06/28/ibm-watson-ai-healthcare/. Accessed: July 26, 2017.

[12]"Mercedes-Benz is connecting the Amazon Echo and Google Home to all its new cars" https://www.theverge.com/2017/4/21/15385232/mercedes-benz-amazon-echo-alexa-google-home. Accessed: July 26, 2017.

4.5   Self-Healing

Just like self-configuring, we need to separate self-healing into two categories, mission critical and the digital platform. Same as self-configuring, the challenges for cognitive life brands will be moving into this mission-critical area. One new essential self-healing aspect is how much easier it is to maintain an electric vehicle versus the combustion engine. We don't want to get into the debate around how fast the electric vehicle market will grow or how much cleaner electric vehicles are than combustion engines, we want to focus on the maintenance side only. The rumors are that new entrants into the automotive and mobility markets will do so with electric vehicles, especially in China, but also by global tier 1 suppliers.

Self-healing is probably one of the main reasons why the simplicity of building and maintaining an electric vehicle can be understood just by comparing the number of moving parts in the drivetrains. An electric motor has an estimated half a dozen parts and the entire drivetrain in the worst case less than 100 moving parts. In comparison, the combustion engine has several hundreds of moving parts and 1000+ on higher-performance engines.[13] The biggest savings in parts are in the engine and gearbox. Also the number of ECUs goes down as well, almost in relation to the number of parts. Still lots of electronics in the vehicle, but from the physical maintenance and healing standpoint a lot easier. This will also effect the organizations within these companies, the focus on quality versus warranty will be quite different, as we stated before even to the point when companies will do away with warranty departments. The main component in an electric vehicle is the battery. Certainly, it will have to be warrantied, but just like most electronic devices, it will need to be replaced overtime anyway.

The digital platform will use the common tools and methods already in place for supporting their smartphone platforms online and remotely. An example was Amazon when it released the Kindle Fire.[14] It came with free online support when the user had trouble with the device. Now that's thinking about the customer experience. Commercial telematics has also used this approach, so we will see it being adopted even more in the future. The blockchain examples showed as well how to deal with the trust and transparency issues during remote support and updates. As vehicles become more and more self-configuring, the transparency and ability to update and maintain these types of configurations will become increasingly critical. Customer will want to know they are safe and all the software, both mission-critical and the digital platform, are up-to-date. The digital industry already has the experiences to meet these challenges. Just think, when was the last time someone actively requested support for a Mac.

---

[13]"Will Electric Cars Require More Maintenance?" http://auto.howstuffworks.com/will-electric-cars-require-more-maintenance.htm. Accessed: July 26, 2017.

[14]"Mayday: Get Help on Your Fire Tablet" https://www.amazon.com/gp/help/customer/display.html?nodeId=201540070. Accessed: July 26, 2017.

4.6   Self-Driving

Audi just announced their A8 to be the first production vehicle with level 3 driving automation. It can only be used in special parts of Germany that have regulations in place for driving this type of vehicle.[15] A few OEMs are making announcements about level 4 of driving automation by 2021. Both insurance and regulations will be roadblocks for the industry in the short term, but if we assume these issues will be addressed, where will we really start to see new types of autonomous vehicles? For more details related to the technical, legal and social challenges of autonomous driving we refer to [4].

We discussed earlier how controlled environments might be the best place to showcase the technology, but let us look more at the mobility industry to understand the effect based on autonomous vehicles. Cities are pursuing autonomous shuttles with announcements from Local Motors[16] and Navya[17] with both launching in the last 2 years. The concepts of shared fleets integrated with public transportation will effect urban mobility as well. General Motors has also made announcements about starting in the ridesharing market before they enter the personal market.[18]

In the commercial industry, both local and long-haul deliveries could be areas dramatically affected and fleet insurance possibly easier to manage. There is also the effect on truck design if we don't have to accommodate a driver, a bit different than personal mobility. What really is exciting is to see companies crossing traditional industry borders and entering the autonomous race, with carriers and retail giants like Alibaba in China and Apple, Google in the US. The types of opportunities that are available based more on a digital focus can really start to affect the industry as well.

We have already talked about Uber, Ford and car2go, so let's look at others in the industry and their approaches to mobility. Lyft, a direct competitor to Uber, has investments from General Motors. Turo is putting a new twist on car rentals by having personal owners sign up to have others rent their vehicles.[19] Zipcar, which is owned by Avis, is carsharing with a business model twist of monthly membership.[20]

---

[15]"The new Audi A8 luxury sedan is a high-tech beast that can drive itself" https://www.theverge.com/2017/7/11/15952510/audi-a8-level-3-autonomous-driving-self-parking. Accessed: July 26, 2017.

[16]"Inside the Local Motors lab where 3D-printed autonomous buses are made" https://www.theverge.com/2016/6/17/11962776/local-motors-olli-3d-printed-autonomous-bus-photos. Accessed: July 26, 2017.

[17]"Las Vegas launches the first electric autonomous shuttle on U.S. public roads" https://techcrunch.com/2017/01/11/las-vegas-launches-the-first-electric-autonomous-shuttle-on-u-s-public-roads/. Accessed: July 26, 2017.

[18]"Self-driving car timeline for 11 top automakers" https://venturebeat.com/2017/06/04/self-driving-car-timeline-for-11-top-automakers/. Accessed: July 26, 2017.

[19]"Rent Better Cars: Choose from thousands of unique cars for rent by local hosts" http://turo.com. Accessed: July 26, 2017.

[20]"Own the Trip, Not the Car" http://www.zipcar.com. Accessed: July 26, 2017.

The most interesting part of this business model is that Avis as the service provider owns the asset, not like Uber, and is basically charging the user a monthly fee and usage for the asset. But all these approaches to mobility are neither established nor a global one-app-fits-all approach.[21]

Carpooling combined with ridesharing is popular in India with ShareACar.[22] BMW has invested in carpooling as well with Scoop, where the focus is on using work colleagues and neighbors for the network of riders.[23] Volkswagen has invested in Gett, a global ridesharing company with a better focus on the driver.[24] Rental companies are getting into the mix as well with Enterprise launching a carsharing service with focus on communities and schools.[25] Toyota is even helping out Uber with the idea that their financial arm can help with leasing and getting more drivers into Toyota vehicles.[26] Fractional vehicle ownership seems to be big in the high-end market with Dreamshare, so we can use different vehicles when we want based on the ownership portion we have. These are just a few examples to show the dynamics currently in the market, and where it may end up in 5–10 years.

Autonomous self-driving combined with mobility could use blockchain solutions for payments and the dynamic contracts. A customer in a carsharing network could use authentication from the solution to gain access to a shared vehicle, unlock the door, and start the engine. Blockchains could handle the contract that arranges the amount of time the vehicle and personalized services related to it are used. All this with a digital vehicle platform that does not need special hardware installed to implement it.

4.7   Self-Socializing

In Sect. 1.7, we introduced e-motion as the e-business way of talking about the new emotional connection between brands and customers that's moving beyond physical and becoming more digital with a focus on the customer experience and creativity with a human touch. What will be interesting to watch is how far the industry takes the concept of human touch, and the challenges they will face doing it. Google and Facebook are already dealing with it today, mainly because some merchants really want that kind of e-motion information because of the types of products they have

[21]"Why Uber's Losing to the Locals in Asia" https://www-bloomberg-com.cdn.ampproject.org/c/s/www.bloomberg.com/amp/view/articles/2017-07-27/why-uber-s-losing-to-local-rivals-in-southeast-asia. Accessed: 27 July 2017.

[22]"Connecting verified users for a comfortable and affordable commute" http://shareacar.co. Accessed: July 26, 2017.

[23]"Meet your new commute" http://www.takescoop.com. Accessed: July 26, 2017.

[24]"Happy drivers means happy riders" http://gett.com. Accessed: July 26, 2017.

[25]"Reserve, Ride, Return" https://www.enterprisecarshare.com/us/en/home.html. Accessed: July 26, 2017.

[26]"Toyota and Uber to Explore Ridesharing Collaboration" http://corporatenews.pressroom.toyota.com/releases/toyota-uber-ridesharing-collaboration-may-24.htm. Accessed: July 26, 2017.

or how the situation may drive a certain type of product. In these cases we are just giving up our GPS data such as,

- where we go,
- how often we are there,
- the type of driving to get there,
- and when we go.

We gave the example of a cognitive vehicle learning the driver's daily routes and behaviors so we could suggest a cup of coffee on the way to work. But what are the other common examples in our life where we need recommendations? Let's think beyond reminders like calendar events, but more in terms of situations like it's time to order more milk, it's time for a glucose shot, remember to get some flowers for the wife that she loves. Or beef stew would be a recommendation based on the available ingredients in the refrigerator for dinner tonight. This again gets back to the creepy line, how comfortable will people really be with this, what will be their e-motion? None of this is new. Most people are already giving this information to Google and their smartphones. The concern is not just the privacy, but the trust in who holds the information as well.

Now start to consider the CPA and the questions it may ask, the potential there is unlimited. Some of the core CPA functions related to driving and merchants are where we like to eat, where we like to stay, where we shop, where we get gas and more. How easy or difficult will it be to configure a CPA like this? When will we want certain recommendations, when will we not want them? This is the more difficult puzzle to solve and related to e-motion, but certainly an area to watch as the industry moves forward. On the first page of the introduction, we asked the question:

► How comfortable are you when your surroundings are more intelligent than you?

Think about the computer-animated film "Inside Out" released by Walt Disney Pictures in 2015 where five personified emotions set in the mind of a young girl are trying to lead her through life.[27] Now reflect on the six e-motion CPAs in Fig. 4.2 that surround us and lead us through life. A deep relationship definitely is the outcome between humans and CPAs. Remember the kid's reaction (see Sect. 1.7):

► Why is it right to love the teddy bear and not to love my cognitive personal assistant?

In Fig. 1.30, we illustrated the six universal emotions, which are distinctly associated with facial expressions for all human beings. It is also known that these emotions can trigger sensations in our bodies—representing 'My Life.' The mind-body connection is biological, which means we all have the same bodily sensations

---

[27]"Inside Out" http://movies.disney.com/inside-out. Accessed: July 26, 2017.

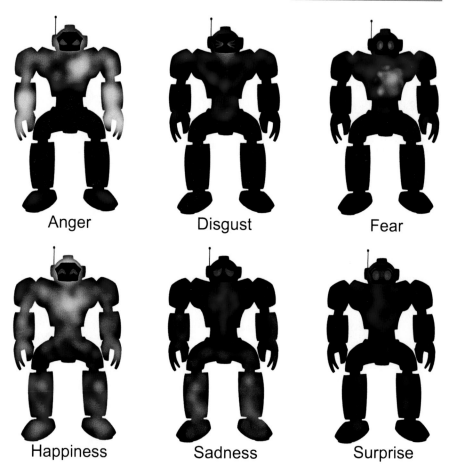

Fig. 4.2 The body maps of the CPA as a service robot reveal the areas in the body where certain sensations may increase (warm colors) or decrease (cool colors) for a given e-motion

associated with the six emotions regardless of culture or language. How will our future CPA as a service robot look like when providing full e-motion (see Fig. 4.2)?

What brands may be behind these e-motion bodies? If we have an iPhone and a Mac, it's not a big leap to think about the Apple TV, an Apple Watch, an Apple Car and Siri as the e-motion CPA. Very similar thinking can be applied to Google, Samsung, Amazon or any other new market entrant. So it seems we will have to pick one brand and CPA, once we pick, how easily can we change?

What will we like and what will we not like? Even if we don't like everything, a digital marital union is evolving. How will the new digital generations cope with this when they love Facebook and Snapchat 1 week and the next week the love changes to the new trendy app? Or is it not ridiculous to hear an Uber user saying "again, I have to wait two more minutes for the Uber car to pick me up, I will use Grab next time." As described in all self-enabling themes, all the personal transactions, learned information and contracts are captured in the CPA. The consequences of such a governance model suffered by all ecosystems are:

Digital divorce? … difficult and painful, but not impossible to manage.

Let's just say we want my cognitive autoMOBILE life without all the strings attached. Like companies and businesses, industry leaders change just like relationships change, so what is an ecosystem today may not be the ecosystem of tomorrow. Tesla and Uber tackled big roadblocks in big traditional industries, and watching more companies break down the other traditional barriers and innovations will be fun to watch. As long as a brand helps me, improves my life, and creates a personalized customer experience, it is on the way to creating my cognitive autoMOBILE life.

References

1. Boesche KV, Rataj D (2016) Zivil- und datenschutzrechtliche Zuordnung von Daten vernetzter Elektrokrafahrzeuge. Begleit- und Wirkungsforschung Schaufenster Elektromobilität, Frankfurt am Main
2. Chua CK, Leong KF (2017) 3D printing and additive manufacturing: principles and applications, 5th edn. World Scientific, Singapore
3. Drescher D (2017) Blockchain basics: a non-technical introduction in 25 steps. Apress, Frankfurt am Main
4. Maurer M, Gerdes JC, Lenz B, Winner H (eds) (2016) Autonomous driving: technical, legal and social aspects. Springer, Berlin
5. Mohr D, Kaas H-W, Gao P, Wee D, Möller T (January 2016) Automotive revolution – perspective towards 2030: How the convergence of disruptive technology-driven trends could transform the auto industry. McKinsey Report, Advanced Industries. http://www.mckinsey.com/industries/automotive-and-assembly/our-insights/disruptive-trends-that-will-transform-the-auto-industry. Accessed 26 July 2017
6. Roßnagel A (2014) Fahrzeugdaten – wer darf über sie entscheiden? SVR Straßenverkehrsrecht, Zeitschrift für die Praxis des Verkehrsjuristen 2014, Heft 8. Nomos Verlagsgesellschaft, Baden-Baden
7. Srivastava AN, Han J (2012) Machine learning and knowledge discovery for engineering systems health management. CRC, Boca Raton, FL
8. Swan M (2015) Blockchain: blueprint for a new economy. O'Reilly Media, Sebastopol, CA

Index

© Springer-Verlag GmbH Germany 2017
S. Wedeniwski, S. Perun, *My Cognitive autoMOBILE Life*,
https://doi.org/10.1007/978-3-662-54677-2